THE INTERNATIONAL ENCYCLOPEDIA
OF PHYSICAL CHEMISTRY AND CHEMICAL PHYSICS

Topic 15. EQUILIBRIUM PROPERTIES OF ELECTROLYTE SOLUTIONS

EDITOR: R. A. ROBINSON

Volume 3

IONIC EQUILIBRIA

BY

J. E. PRUE

IONIC EQUILIBRIA

BY

J. E. PRUE

DEPARTMENT OF CHEMISTRY,
THE UNIVERSITY, READING

PERGAMON PRESS
OXFORD · LONDON · EDINBURGH · NEW YORK
TORONTO · PARIS · FRANKFURT

Pergamon Press Ltd., Headington Hill Hall, Oxford
4 & 5 Fitzroy Square, London W.1

Pergamon Press (Scotland) Ltd., 2 & 3 Teviot Place, Edinburgh 1

Pergamon Press Inc., 44-01 21st Street, Long Island City, New York 11101

Pergamon of Canada, Ltd., 6 Adelaide Street East, Toronto, Ontario

Pergamon Press, S.A.R.L., 24 rue des Écoles, Paris 5e

Pergamon Press GmbH, Kaiserstrasse 75, Frankfurt-am-Main

Copyright © 1966
Pergamon Press Ltd.

First edition 1966

Library of Congress Catalog Card No. 65–19995

PRINTED IN GREAT BRITAIN BY BELL AND BAIN LTD., GLASGOW

2365/66

THE INTERNATIONAL ENCYCLOPEDIA
OF PHYSICAL CHEMISTRY AND CHEMICAL PHYSICS

Editors-in-Chief

E. A. GUGGENHEIM   J. E. MAYER
READING   LA JOLLA

F. C. TOMPKINS
LONDON

Chairman of the Editorial Advisory Group

ROBERT MAXWELL
PUBLISHER AT PERGAMON PRESS

List of Topics and Editors

1.	Mathematical Techniques	H. JONES, *London*
2.	Classical and Quantum Mechanics	R. McWEENY, *Keele*
3.	Electronic Structure of Atoms	C. A. HUTCHISON, JR., *Chicago*
4.	Molecular Binding	J. W. LINNETT, *Cambridge*
5.	Molecular Properties	
	(a) Electronic	J. W. LINNETT, *Cambridge*
	(b) Non-electronic	N. SHEPPARD, *East Anglia*
6.	Kinetic Theory of Gases	E. A. GUGGENHEIM, *Reading*
7.	Classical Thermodynamics	D. H. EVERETT, *Bristol*
8.	Statistical Mechanics	J. E. MAYER, *La Jolla*
9.	Transport Phenomena	J. C. McCOUBREY, *Birmingham*
10.	The Fluid State	J. S. ROWLINSON, *London*
11.	The Ideal Crystalline State	M. BLACKMAN, *London*
12.	Imperfections in Solids	Editor to be appointed
13.	Mixtures, Solutions, Chemical and Phase Equilibria	M. L. McGLASHAN, *Exeter*
14.	Properties of Interfaces	D. H. EVERETT, *Bristol*
15.	Equilibrium Properties of Electrolyte Solutions	R. A. ROBINSON, *Washington, D.C.*
16.	Transport Properties of Electrolytes	R. H. STOKES, *Armidale*
17.	Macromolecules	C. E. H. BAWN, *Liverpool*
18.	Dielectric and Magnetic Properties	J. W. STOUT, *Chicago*
19.	Gas Kinetics	A. F. TROTMAN-DICKENSON, *Aberystwyth*
20.	Solution Kinetics	R. M. NOYES, *Eugene*
21.	Solid and Surface Kinetics	F. C. TOMPKINS, *London*
22.	Radiation Chemistry	R. S. LIVINGSTON, *Minneapolis*

THE INTERNATIONAL ENCYCLOPEDIA
OF PHYSICAL CHEMISTRY AND CHEMICAL PHYSICS

Members of the Honorary Editorial Advisory Board

J. N. AGAR, *Cambridge*
R. M. BARRER, *London*
C. E. H. BAWN, *Liverpool*
N. S. BAYLISS, *Western Australia*
R. P. BELL, *Oxford*
C. J. F. BÖTTCHER, *Leyden*
F. P. BOWDEN, *Cambridge*
G. M. BURNETT, *Aberdeen*
J. A. V. BUTLER, *London*
C. A. COULSON, *Oxford*
J. S. COURTNEY-PRATT, *New Jersey*
D. P. CRAIG, *London*
F. S. DAINTON, *Nottingham*
C. W. DAVIES, *London*
B. V. DERJAGUIN, *Moscow*
M. J. S. DEWAR, *Texas (Austin)*
G. DUYCKAERTS, *Liège*
D. D. ELEY, *Nottingham*
H. EYRING, *Utah*
P. J. FLORY, *Mellon Institute*
R. M. FUOSS, *Yale*
P. A. GIGUÈRE, *Laval*
W. GROTH, *Bonn*
J. GUÉRON, *Brussels*
C. KEMBALL, *Queen's, Belfast*
J. A. A. KETELAAR, *Amsterdam*
G. B. KISTIAKOWSKY, *Harvard*
H. C. LONGUET-HIGGINS, *Cambridge*
R. C. LORD, *Massachusetts Institute of Technology*
M. MAGAT, *Paris*

R. MECKE, *Freiburg*
SIR HARRY MELVILLE, *S.R.C., London*
S. MIZUSHIMA, *Tokyo*
R. S. MULLIKEN, *Chicago*
R. G. W. NORRISH, *Cambridge*
R. S. NYHOLM, *London*
J. T. G. OVERBEEK, *Utrecht*
K. S. PITZER, *Rice University, Houston*
J. R. PLATT, *Chicago*
G. PORTER, *The Royal Institution (London)*
I. PRIGOGINE, *Brussels (Free University)*
R. E. RICHARDS, *Oxford*
SIR ERIC RIDEAL, *London*
J. MONTEATH ROBERTSON, *Glasgow*
E. G. ROCHOW, *Harvard*
G. SCATCHARD, *Massachusetts Institute of Technology*
GLENN T. SEABORG, *California (Berkeley)*
N. SHEPPARD, *East Anglia*
R. SMOLUCHOWSKI, *Princeton*
H. STAMMREICH, *São Paulo*
SIR HUGH TAYLOR, *Princeton*
H. G. THODE, *McMaster*
H. W. THOMPSON, *Oxford*
D. TURNBULL, *G.E., Schenectady*
H. C. UREY, *California (La Jolla)*
E. J. W. VERWEY, *Philips, Eindhoven*
B. VODAR, *Laboratoire de Bellevue, France*
M. KENT WILSON, *Tufts*
LORD WYNNE-JONES, *Newcastle-upon-Tyne*

CONTENTS

		PAGE
INTRODUCTION		ix
PREFACE		xi
SYMBOLS		xiii
ELECTROCHEMICAL CONSTANTS USED IN CALCULATIONS		xiv

CHAPTER 1	INTRODUCTION	1
1.1	Preliminary	1
1.2	Historical	1
1.3	Equilibrium constants and quotients	4
1.4	Terminology	6
1.5	Arrangement of subsequent chapters	7

CHAPTER 2	OPTICAL ABSORPTION, RAMAN AND NUCLEAR MAGNETIC RESONANCE SPECTROSCOPY	9
2.1	Introduction	9
2.2	Simple acid–base equilibria	12
2.3	Use of acid–base indicators	12
2.4	Simple complex ion equilibria	14
2.5	Overlapping complex ion equilibria	18
2.6	Raman spectroscopy	19
2.7	Nuclear magnetic resonance spectroscopy	22

CHAPTER 3	CONDUCTANCE	25
3.1	Introduction	25
3.2	Weak electrolytes	25
3.3	Strongly dissociated electrolytes	30
3.4	Mixed electrolytes	34

CHAPTER 4	ELECTROCHEMICAL CELLS	36
4.1	Introduction	36
4.2	Equilibrium constants of electron-transfer reactions	37
4.3	Study of protolytic equilibria by measurement of $[H^+]\gamma_H\gamma_{Cl}$	38
4.4	Thermodynamic stability constants of metal ion complexes from cells	41
4.5	Use of swamping ionic media	44
4.6	The calculation of successive stability constants	46
4.7	Polarographic measurement of cation concentration	48

CHAPTER 5	SOLUBILITY AND DISTRIBUTION	51
5.1	Solubility	51
5.2	Distribution method	55
5.3	Ion-exchange method	58

CONTENTS

			PAGE
CHAPTER 6		COLLIGATIVE PROPERTIES	61
	6.1	Introduction	61
	6.2	Cryoscopy	61
	6.3	Isopiestic measurements	65
CHAPTER 7		RATES OF REACTION	69
	7.1	Introduction	69
	7.2	Protolytic equilibria	70
	7.3	Complex ion equilibria	71
CHAPTER 8		RELAXATION SPECTROMETRY	74
	8.1	Introduction	74
	8.2	Principles of method	74
	8.3	Experimental techniques	76
	8.4	Results	77
CHAPTER 9		ACIDITY CONSTANTS	82
	9.1	Introduction	82
	9.2	Substituent effects on protolytic equilibria	85
	9.3	pK_a values of oxyacids	87
CHAPTER 10		STABILITY CONSTANTS	90
	10.1	Introduction	90
	10.2	The electrostatic model for complex ion equilibria	91
	10.3	Ionic hydration	95
	10.4	Step stability constants	96
	10.5	d^0 cations	97
	10.6	d^{10} cations	100
	10.7	d^n cations	102
	10.8	Chelate stability	104
CHAPTER 11		NON-AQUEOUS AND MIXED SOLVENTS	107
	11.1	Simple electrostatic model	107
	11.2	Specific solvation effects	109
	11.3	Tetra-alkylammonium salts	111
INDEX			113

INTRODUCTION

The International Encyclopedia of Physical Chemistry and Chemical Physics is a comprehensive and modern account of all aspects of the domain of science between chemistry and physics, and is written primarily for the graduate and research worker. The Editors-in-Chief, Professor E. A. Guggenheim, Professor J. E. Mayer and Professor F. C. Tompkins, have grouped the subject matter in some twenty groups (General Topics), each having its own editor. The complete work consists of about one hundred volumes, each volume being restricted to around two hundred pages and having a large measure of independence. Particular importance has been given to the exposition of the fundamental bases of each topic and to the development of the theoretical aspects; experimental details of an essentially practical nature are not emphasized although the theoretical background of techniques and procedures is fully developed.

The Encyclopedia is written throughout in English and the recommendations of the International Union of Pure and Applied Chemistry on notation and cognate matters in physical chemistry are adopted. Abbreviations for names of journals are in accordance with *The World List of Scientific Periodicals*.

PREFACE

IONIC equilibria are studied by two rather different kinds of chemist; one can call them electrolyte solution chemists and complex ion chemists respectively. The former are interested in the precise study of simple equilibria at low ionic concentrations, the latter in less precise data for a wide range of what are often complicated equilibria. It is my general impression that each class takes too little cognisance of the work of the other class. Drastic approximations are sometimes necessary, particularly in the second type of work. There is no harm in this, provided the nature of the approximations is clearly stated, which is all too seldom done in the literature.

The various methods of studying ionic equilibria are capable of rigorous discussion in simple language without the obscuration of principles by tedious algebraic detail. This I have tried to avoid. Nevertheless it is only by careful examination of the successive stages of numerical calculations for selected cases that it is possible to lay bare the importance of assumptions and the likely sources of error, and to assess the physical significance of the constants that emerge. Fitting equilibrium constants to experimental data can easily degenerate into meaningless numerology.

There is a chapter on relaxation spectrometry; although it is a branch of kinetics the information it provides on equilibria is not obtainable in other ways.

The equilibrium constants themselves show some simple and interesting patterns which are discussed in the final chapters, even though much of the detail is puzzling and quantitative prediction usually impossible. Again it seems to me that the two classes of chemist take too little cognisance of one another's contributions.

No attempt has been made to provide a bibliography of the vast amount of work, good, bad and indifferent, which has been published on ionic equilibria. There are two topics on which much more could have been written. The first is the analysis of results where successive stages in the formation of complex ions or in the dissociation of a polybasic acid overlap, and the second is in the finer details of theoretical discussions of variations of equilibrium constants or quotients with medium, temperature and substituents. An extended treatment of either of these topics would not only have made the book too long, but would also have distorted the general balance which I had in mind.

The topic editor (Dr. R. A. Robinson) and Mr. R. P. Bell, Professor E. J. King and Dr. M. L. McGlashan kindly read the entire draft manuscript. I also had the benefit of comments on several chapters from Dr. A. K. Covington, Professor E. A. Guggenheim, Dr. P. C. H. Mitchell and Professor R. H. Stokes. I am most grateful to all of them; I usually agreed with their criticisms, but the responsibility for the final outcome is, of course, my own.

J. E. PRUE

Reading.
September, 1964

SYMBOLS

A	constant of Debye–Hückel equation for $\log f$ (A' for $\log \gamma$); optical absorbance
a	closest distance of approach of ions in an ion-pair
a_0	coefficient of ion-size term in Debye–Hückel equation for f (a_0' for γ)
B	ionic specific interaction coefficient in equation for $\log f$ (B' for $\log \gamma$)
c	molar concentration
d	closest distance of approach of *free* ions (association distance)
E	electromotive force; parameter in Fuoss–Onsager conductance equation
e	proton charge
F	Faraday
F	$\Lambda/\alpha\Lambda_0$
f	activity coefficient on molarity scale
I	ionic strength on molarity scale (I' on molality scale); light or sound intensity
i	current
J	parameter in Fuoss–Onsager conductance theory
K	equilibrium constant
k	Boltzmann's constant
k	rate constant
L	a ligand
l	length
ln, log	logarithm to base e and 10 respectively
M	abbreviation for mole l^{-1}
N	Avogadro's number
$\bar{n}$	ligand number
P	solubility product; partition coefficient
p	pressure
p	an operator denoting $-\log$
Q	equilibrium quotient; absorption coefficient for sound; partition function
R	gas constant
R	buffer ratio; distribution ratio

r	distance between two ions
s	$z_+ \mid z_- \mid e^2/\epsilon kT$
T	absolute temperature; optical transmittance
V	volume
z_i	charge number on ion i (negative for a negative ion)
α	degree of dissociation
α_D	constant of Debye–Hückel equation for $\ln f$ (α'_D for $\ln \gamma$)
β	overall stability quotient
γ	activity coefficient on molality scale
δ	chemical shift in nuclear magnetic resonance
ϵ	molar absorptivity; dielectric constant
θ	coefficient of the relaxation term in the theory of conductivity
κ	Debye–Hückel characteristic reciprocal distance; absorption (extinction) coefficient of a solution; compressibility
Λ	molar conductance
λ	cryoscopic constant; wavelength
ν	frequency
$\Pi(f)$	product of activity coefficients
ρ	d/a_0 (ρ' on molality scale)
σ	coefficient of the electrophoretic term in the theory of conductivity
τ	relaxation time
ϕ	molal osmotic coefficient
[A]	concentration of species A

Electrochemical constants used in calculations

Values are for water at 25°C unless otherwise noted.

A $0\cdot5115$ mole$^{-\frac{1}{2}}$l$^{\frac{1}{2}}$

A' $0\cdot5107$ mole$^{-\frac{1}{2}}$kg$^{\frac{1}{2}}$ ($0\cdot4917$ mole$^{-\frac{1}{2}}$kg$^{\frac{1}{2}}$ at 0°C)

a_0 $3\cdot039$ Å mole$^{\frac{1}{2}}$l$^{-\frac{1}{2}}$

a'_0 $3\cdot043$ Å mole$^{\frac{1}{2}}$kg$^{-\frac{1}{2}}$ ($3\cdot079$ Å mole$^{\frac{1}{2}}$kg$^{-\frac{1}{2}}$ at 0°C)

$(RT \ln 10)/F$ $59\cdot160$ mV

θ $0\cdot2300$ mole$^{-\frac{1}{2}}$l$^{\frac{1}{2}}$

σ $60\cdot65$ mole$^{-\frac{1}{2}}$l$^{\frac{1}{2}}$ int. Ω^{-1} mole^{-1}cm^2

$e^2/\epsilon kT$ $7\cdot135$ Å ($6\cdot935$ Å at 0°C)

CHAPTER 1

INTRODUCTION

1.1. Preliminary

Volumes 1 and 2 on the general topic " Equilibrium properties of electrolyte solutions " will be substantially concerned with the success of models that take no account of specific " chemical " equilibria in the solutions. This volume, on the other hand, deals with solutions in which the idea of ions and molecules of different natures and structures being in chemical equilibrium with one another is paramount. The distinction is not a sharp one and it is sometimes a matter of convenience rather than necessity to treat, for example, two oppositely charged ions which are close together as a separate chemical species. There are four major classes of ionic equilibria—protolytic, complex ion, solubility and oxidation–reduction. Typical examples of each class are

$$H^+ + NH_3 \rightleftharpoons NH_4^+$$
$$Ag^+ + 2NH_3 \rightleftharpoons Ag(NH_3)_2^+$$
$$Ag^+ + Cl^- \rightleftharpoons AgCl(s)$$
$$2Ag^+ + C_6H_4(OH)_2 \rightleftharpoons 2Ag + C_6H_4O_2 + 2H^+$$

These equations leave open the extent of solvent interaction with the various species. There is little doubt that in water, for example, ligands such as NH_3 and Cl^- are in competition with the solvent molecules for the cation. Equilibrium constants of ionic equilibria are determined by physical chemists primarily concerned with the interpretation and correlation of the various properties of solutions, by inorganic and analytical chemists interested in the nature and stability of complexes, by biochemists interested in the role of such complexes in biochemical reactions, and by physical organic chemists concerned with the nature of the reacting species in organic reactions, particularly in non-aqueous solvents.

1.2. Historical

Almost a century ago the quantitative study of the chemical equilibrium attained in mixtures of ethyl alcohol and acetic acid led to the classical form of the law of chemical equilibrium, usually known as the law of mass action (for the early history see Guggenheim[1]). In the first volume of the *Zeitschrift für physikalische Chemie*, Arrhenius[2] proposed that electrolytes are dissociated into ions to a degree which could be determined from the conductance or colligative properties of the solution, and in the second volume Ostwald[3] showed that the behaviour of solutions of carboxylic acids conformed to the law of mass action. If, following Arrhenius, the degree of dissociation α of an electrolyte at concentration c is given by the ratio of the molar conductance Λ to the conductance at infinite dilution Λ_0, combination of

TABLE 1.1
Classical equilibrium constants for acetic acid and sodium acetate

	HOAc ($\Lambda_0 = 390\cdot7$ int. Ω^{-1} cm^2 mole^{-1})		NaOAc ($\Lambda_0 = 91\cdot0$ int. Ω^{-1} cm^2 mole^{-1})	
c mole l^{-1}	Λ int. Ω^{-1} cm^2 mole^{-1}	$10^5 K$ mole l^{-1}	Λ int. Ω^{-1} cm^2 mole^{-1}	K mole l^{-1}
0·001	48·65	1·76	88·5	0·034
0·005	22·83	1·82	85·72	0·078
0·01	16·24	1·80	83·76	0·106
0·02	11·57	1·80	81·24	0·152
0·05	7·36	1·80	76·92	0·230
0·10	5·20	1·79	72·80	0·320

the relation $\Lambda/\Lambda_0 = \alpha$ with the law of chemical equilibrium $\alpha^2 c/(1-\alpha) = K$ gives $K = (\Lambda/\Lambda_0)^2 c/(1 - \Lambda/\Lambda_0)$. Table 1.1 shows that the classical equilibrium constant calculated in this way is constant for acetic acid but not for sodium acetate. Discrepancies of the latter kind and the disagreement between apparent degrees of dissociation determined for the same solution from the conductance and from a colligative property constituted one of the so-called " anomalies of strong electrolytes ", which was not quantitatively resolved until the nineteen-twenties for dilute solutions, and has not yet been quantitatively resolved for concentrated solutions. Nevertheless its existence did not deter experimentalists from determining equilibrium constants for the dissociation of a wide range of weak electrolytes. Not only were protolytic equilibria studied by conductance and kinetic methods, but the introduction by Nernst[4] of the e.m.f. method for determining metal ion concentrations was widely exploited at the beginning of the present century by Bodländer, Abegg and others in the study of the complex ion equilibria of Ag^+, Cu^+, and Hg^{2+}. Distribution coefficient and solubility methods were also used. The history of methods of studying ionic equilibrium has been reviewed in an article by Sillén[5] where full reference to the early work can be found.

What is in principle required for the determination of the equilibrium constant of an ionic reaction is a method of determining the concentration, or some quantity closely related to it, of one of the participants in the equilibrium (the concentrations of the others can then usually be calculated from their total stoichiometric concentrations and the stoichiometry of the reaction). The most direct method, that of chemical analysis, can only be used in cases of non-labile equilibria. The slow formation and decomposition of (inner-sphere) chromic complexes was exploited by N. Bjerrum[6] in 1915, who was able to measure independently the six successive equilibrium constants for the formation of $Cr(SCN)_6^{3-}$ from Cr^{3+} and SCN^- after having developed procedures for the analysis of a solution for each complex. Equilibria involving chromic complexes are still studied[7] by analytical procedures, the separation of complexes by ion-exchange columns prior to analysis being a useful aid.

Little progress in the study of ionic equilibria involving other than weak electrolytes was made until the effects of long-range interionic forces on the properties of solutions had been appreciated[8] and quantitatively expressed,[9] although the notion of maintaining constant effects other than those due to the equilibrium under study by adding a swamping excess of an inert salt seems to have been suggested by Bodländer (who died in 1904) to his student Grossmann. Grossmann[10] published in 1905 the results of a study of the complex formation between Hg^{2+} and SCN^- in which potassium nitrate was added to maintain $[K^+] = 1M$. Later Brönsted and Pedersen[11] showed that the equilibrium quotient for the reaction $Fe^{3+} + I^- \rightleftarrows Fe^{2+} + \frac{1}{2}I_2$ was constant in a swamping medium of $1 \cdot 65M$ KCl $+ 0 \cdot 1M$ HCl. The spectacular success of the Debye–Hückel equations together with the assumption of complete dissociation in quantitatively describing the thermodynamic properties of dilute aqueous solutions of alkali and alkaline-earth metal salts was followed by a period when all salts, apart from a few curiosities like mercuric chloride, were regarded as completely dissociated in water. This attitude persisted long after discrepancies between experimental results and theoretical predictions for thermodynamic[12] and conductimetric[13] properties of salts of multiply charged and particularly transition metal cations in water, and measurements in non-aqueous solvents, had been interpreted as due to incomplete dissociation into free ions obeying the theoretical formulae. The spectrophotometric method used was questioned[14] when in 1949[15] specific salt effects at constant ionic strength on the equilibrium $Fe^{3+} + OH^- \rightleftarrows FeOH^{2+}$ were observed. A satisfactory explanation of the effects can be given[16] in terms of the association of some of the ferric ions with perchlorate ions. Of course, there is no reason why, in principle, a theory of electrolyte solutions should not contain "built-in" provision for short-range specific interactions between ions not only of an ionic but also of a covalent nature without the explicit introduction of the concept of incomplete dissociation. So far, however, the simplest, most convenient and most generally applicable procedure for coping with the effects of strong short-range interactions between oppositely charged ions is to absorb the details in dissociation constants the theoretical significance of which can be analysed subsequently. In the last two decades the quantitative information available about ionic equilibria has increased rapidly as the result of measurements made by those primarily interested in the nature and relative stabilities of the complexes formed by transition metal cations with a variety of inorganic and organic anions and molecules; the constant ionic medium has been widely, and no doubt sometimes improperly,[17] used. A compilation[18] by three of the foremost workers in the field, J. Bjerrum, G. Schwarzenbach and L. G. Sillén, provides an invaluable source of reference to work on complex ion, protolytic and solubility equilibria.

1.3. Equilibrium constants and quotients

By far the most extensively studied ionic equilibria are those in which a cation combines with an anion or neutral molecule. The thermodynamic equilibrium constant K for a reaction $A^{a+} + B^{b-} \rightleftarrows AB^{(a-b)+}$ is given by

$$K = \frac{[AB^{(a-b)+}]}{[A^{a+}][B^{b-}]} \frac{f_{AB}}{f_A \cdot f_B} = Q \cdot \frac{f_{AB}}{f_A \cdot f_B} \qquad (1.3.1)$$

where [ ] denotes a concentration, f_{AB}, f_A, f_B are activity coefficients usually defined so that $f \rightarrow 1$ as the composition of the medium tends to that of the pure solvent and Q, the classical equilibrium constant, is a quotient which is only constant in so far as the activity coefficient factor remains constant. With a series of overlapping equilibria, a situation often encountered in complex ion formation, a useful and almost necessary simplification is effected by making all measurements in a swamping medium of an " inert " electrolyte and then supposing that the equilibrium equation can be used in the classical form. One of the most popular media is 3M sodium perchlorate because the tendency of both ions to form complex ions is slight. Nevertheless in such a medium there must be specific and unknown interactions between added ions and those of the medium (the average interionic distance in 3M sodium perchlorate is only 6·5 Å) and our present theoretical or experimental knowledge is inadequate to permit extrapolation of quantitative conclusions to media other than those in which the measurements have been made.

In the case of relatively weak electrolytes, especially in the absence of overlapping equilibria, knowledge of the activity coefficient factor is sufficient to permit the calculation of K from values of Q. On rare occasions, if the electrolyte is weak enough and sufficiently accurate measurements can be made, the limiting law of Debye and Hückel[9] can be used for the activity coefficients

$$-\ln f_i = z_i^2 \alpha_D I^{\frac{1}{2}} \text{ or } -\log f_i = z_i^2 A I^{\frac{1}{2}} \qquad (1.3.2)$$

where α_D is a constant of the theory depending on the properties of the solvent and I is the ionic strength defined by $I = \Sigma \frac{1}{2} c_i z_i^2$ where c is the concentration of an ion i of charge number z_i (negative for a negative ion). The activity coefficient factor for the dissociation of an uncharged acid may be in error by as much as 5 per cent if this formula is used even at an ionic strength as low as 0·001M. It is preferable to use the formula

$$-\log f_i = z_i^2 A I^{\frac{1}{2}} / (1 + \rho I^{\frac{1}{2}}) \qquad (1.3.3)$$

where $\rho = d/a_0$, d being the mean ionic diameter* and a_0 a constant of the theory. Unfortunately this has introduced another adjustable parameter, for we have no exact knowledge of d from other experimental data. However, a series of measurements of Q over an ionic strength range can be used to obtain both K and ρ. It is convenient that $AI^{\frac{1}{2}}/\{1 + (\rho + \Delta\rho)I^{\frac{1}{2}}\}$ is closely approximated by $AI^{\frac{1}{2}}/(1 + \rho I^{\frac{1}{2}}) - A\Delta\rho I$ provided that $\Delta\rho$ is small. This makes it possible arbitrarily to fix ρ at approximately the correct magnitude and then write

$$-\log f_i = z_i^2 A I^{\frac{1}{2}} / (1 + \rho I^{\frac{1}{2}}) - B_i I \qquad (1.3.4)$$

* d is used to denote the closest distance of approach of *free* ions so that a can be used for the closest distance of approach of paired ions in an ion-pair (1.4).

which gives in combination with equation 1.3.1

$$\log K = \log Q - \frac{2Az_Az_BI^{\frac{1}{2}}}{1 + \rho I^{\frac{1}{2}}} + BI \qquad (1.3.5)$$

where B is constant. The linear term will in fact take account of a number of effects not incorporated in the Debye–Hückel model. It is often convenient to put $\rho = 1$. From a plot of the first two terms on the right-hand side of equation 1.3.5 against I, K and B (and hence the correct ρ if desired) can be obtained.

Properly[19] equation 1.3.4 should be replaced by

$$-\log f_i = z_i^2 A I^{\frac{1}{2}}/(1 + \rho I^{\frac{1}{2}}) - \Sigma B_{ij} c \qquad (1.3.6)$$

where B_{ij} is a specific interaction coefficient and the summation extends over all ions of opposite charge to the ion of type i. This may introduce additional parameters but such specific interaction coefficients are sometimes known from independent experiments. Clearly the higher the ionic strengths at which measurements are made the more serious the difficulties in obtaining the thermodynamic constant become; theoretical prediction of activity coefficients becomes well-nigh impossible above ionic strengths of about 0·1M in water even in the absence of highly charged ions. The major problem with strongly dissociated electrolytes is, however, not the conversion of values of Q to K, but rather the determination of unambiguous values of Q.

The symbol γ will be used for activity coefficients on the molality scale; correspondingly, the other symbols in equation 1.3.4 will be primed, e.g. A'.

1.4. Terminology

Thermodynamic constants, and often equilibrium quotients, of the type in equation 1.3.1 are known as stability, association, complexity or formation constants. The inverse quantity is known as a dissociation, acidity or ionization constant for an acid, and a dissociation or instability constant for a complex ion.

The usages of the terms ion-pair and complex vary and can cause confusion. The reason is that writers differ in the kind of information they try to convey by the use of the two words. In the commonest usage the term ion-pair is restricted to an associated pair of ions which have a stability constant approximately equal to that to be expected from coulombic attraction alone between the two ions, and in some property behave in a manner distinct from the free constituent ions.

Those who use the term in this sense are making a judgement about the nature of the bonding (often based on their "chemical intuition" rather than on quantitative calculation). They are particularly likely to be correct when the two ions are separated by an interposed layer of solvent or other neutral ligands, e.g. $Co(NH_3)_6^{3+}SO_4^{2-}$. Such a species is alternatively described[20] as an outer-sphere complex by contrast with an inner-sphere complex where there is no such interposed layer, e.g. $Co(NH_3)_5SO_4^+$. Some writers[21] prefer to restrict the term ion-pair to outer-sphere complexes. Others[22,23] distinguish between "internal" "intimate" or "contact" ion-pairs in contrast with "external" or "solvent-separated" pairs. It has been suggested[23] that a distinction be made between "solvent-shared" ion-pairs $A^+OH_2B^-$ and "solvent-separated" ion-pairs $A^-(OH_2)_nB^-$ where $n > 1$. Another usage[24] restricts the term ion-pair to the species $A^+OH_2B^-$ in contrast with a complex A^+B^- or associated ions $A^+(OH_2)_nB^-$. The invention of labels has outrun the ways of distinguishing between species in a significant number of cases. The suggestion of additional ones would not be helpful. It seems simplest and best to use the term complex quite generally and to qualify adjectively when it is both necessary and possible to do so. It will also occasionally be useful to use the term ion-pair in its commonest sense.

1.5. Arrangement of subsequent chapters

Ionic equilibria manifest themselves by their effects on many if not all of the physicochemical properties of a solution. Quantitative study of these effects leads to values of equilibrium constants. In the next eight chapters the most commonly and usefully studied properties are considered in turn and ways of analysing the experimental data are discussed in detail. If physical significance is to attach to an equilibrium constant it is necessary to examine carefully how it depends on assumptions made in obtaining it and whether values of the same constant obtained from different properties agree. Details of experimental techniques are not discussed. The final three chapters examine the systematic patterns shown by ionic equilibrium constants, the degree to which some of these patterns can be related to simple theoretical models, and their dependence on the solvent.

The book is not intended to serve as a source of equilibrium constant values for individual systems. References are usually restricted to reviews and papers which are concerned with some point of principle, contain data which are discussed in detail, or typify the application of a particular method. Information about individual systems is conveniently sought in the compilation "Stability Constants"[18] or in articles in the *Annual Reports of Progress in Chemistry* and the *Annual Reviews of Physical Chemistry*.

REFERENCES

1. GUGGENHEIM, E. A., *J. chem. Educ.*, 1956, **33**, 544.
2. ARRHENIUS, S., *Z. phys. Chem.*, 1887, **1**, 631.
3. OSTWALD, W., *Z. phys. Chem.*, 1888, **2**, 277.
4. NERNST, W., *Z. phys. Chem.*, 1889, **4**, 129.
5. SILLÉN, L. G., *J. Inorg. Nuclear Chem.*, 1958, **8**, 176.
6. BJERRUM, N., *Kgl. danske Videnskab. Selskab Skrifter, mat. nat. Afdel*, 1915, **12**, 4.
7. ESPENSON, J. H. and KING, E. L., *J. phys. Chem.*, 1960, **64**, 380.
8. BJERRUM, N., *7th Internat. Congr. Applied Chem.*, 1909, Section x, p. 58.
9. DEBYE, P. and HÜCKEL, E., *Physik. Z.*, 1923, **24**, 185, 305.
10. GROSSMANN, H., *Z. anorg. Chem.*, 1905, **43**, 356.
11. BRÖNSTED, J. N. and PEDERSEN, K., *Z. phys.*, 1922, **103**, 307.
12. BJERRUM, N., *Kgl. danske Videnskab. Selskab, Mat.-fys. Medd.*, 1926, **7**, No. 9.
13. DAVIES, C. W., *Trans. Faraday Soc.*, 1927, **23**, 351.
14. GORDON, A. R., *Ann. Rev. Phys. Chem.*, 1950, **1**, 71.
15. OLSON, A. R. and SIMONSON, T. R., *J. chem. Phys.*, 1949, **17**, 1322.
16. SYKES, K. W., *J. chem. Soc.*, 1959, 2473.
17. YOUNG, T. F. and JONES, A. C., *Ann. Rev. Phys. Chem.*, 1952, **3**, 286.
18. BJERRUM, J., SCHWARZENBACH, G. and SILLÉN, L.G., "Stability Constants", Parts I and II, *Chem. Soc. Special Publ.*, 1957, No. 6 ; 1958, No. 7
    SILLÉN, L. G. and MARTELL, A. E., *ibid.*, 2nd. ed., 1964, No. 17.
19. GUGGENHEIM E. A. and TURGEON, J. C., *Trans. Faraday Soc.*, 1955, **51**, 747.
20. POSEY, F. A. and TAUBE, H., *J. Amer. chem. Soc.*, 1956, **78**, 15.
21. BASOLO, F. and PEARSON, R. G., *Mechanisms of Inorganic Reactions*, John Wiley, New York, 1958, p. 376.
22. WINSTEIN, S. and ROBINSON, G. C., *J. Amer. chem. Soc.*, 1958, **80**, 169.
23. GRIFFITHS, T. R. and SYMONS, M. C. R., *Mol. Phys.*, 1960, **3**, 90.
24. DUNCAN, J. F., *Disc. Faraday Soc.*, 1957, **24**, 129.

CHAPTER 2

OPTICAL ABSORPTION, RAMAN AND NUCLEAR MAGNETIC RESONANCE SPECTROSCOPY

2.1. Introduction

Many equilibria have been studied by optical absorption methods and the first five sections deal with these alone, and the last two with the more recently developed Raman and nuclear magnetic resonance (n.m.r.) techniques. The ionization of a proton from an aromatic molecule causes pronounced changes in the visible or ultraviolet absorption bands, and the same is true of the attachment of a ligand to a transition metal cation. This, together with the insensitivity of the absorption spectra of ions to the long-range coulombic interactions between ions which affect the thermodynamic and conductance properties, makes the spectrophotometric method useful for the study of both protolytic and complex ion equilibria. The method has developed rapidly since 1945 with the production of a wide range of commercial instruments capable of making measurements not only in the visible but also down to 2000 Å in the ultraviolet.

It is interesting to enquire how intimate the interaction of an anion with a cation needs to be in order to affect the absorption spectrum of the latter. It is necessary to distinguish between two kinds of absorption band. The first kind are bands of low intensity in the visible part of the spectrum due to transitions of electrons between orbitals in incompletely filled d-subshells of the cations; the recent successes of ligand field theory in interpreting such spectra are noteworthy.[1] Inner-sphere complex formation is necessary before these bands are appreciably affected. For example, the molar absorptivity in the visible of the non-labile cation $Cr(H_2O)_6^{3+}$ is independent of concentration or anion; this[2] was one of the earliest pieces of evidence for the complete ionization of these salts. The similarity of the visible absorption spectra[3] of aqueous solutions of transition metal cations such as Ni(II) with those of hydrated salts, e.g. $NiSO_4 \cdot 7H_2O$, provides strong evidence that the ion in solution is in the same immediate environment of six octahedrally co-ordinated water molecules as in the solid hydrate. Figure 2.1 shows how slightly the spectrum of $Ni(H_2O)_6^{2+}$ in the visible and near infrared is affected by its environment. Changes in d–d bands are evidence of inner-sphere complex formation;[4] these changes are particularly marked if the complex formation is accompanied by a change of co-ordination number which changes the symmetry of the ligand field, e.g. formation of blue chloro-complexes with tetrahedral stereochemistry from pink $Co(H_2O)_6^{2+}$. The second kind of absorption band found with transition metal cations usually occurs in the ultraviolet

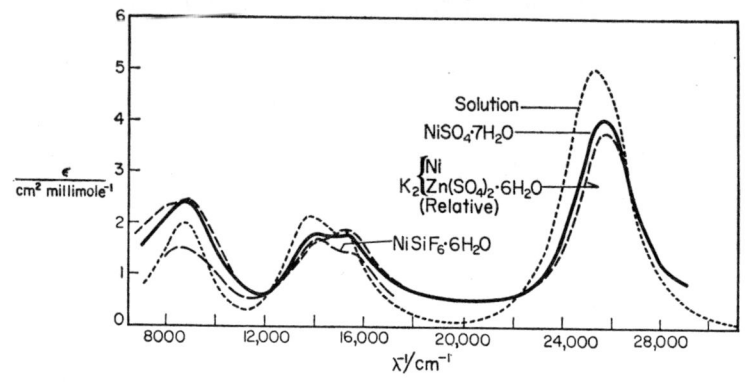

FIG. 2.1. Absorption spectrum of $Ni(H_2O)_6^{2+}$ in different environments. (From reference 3.)

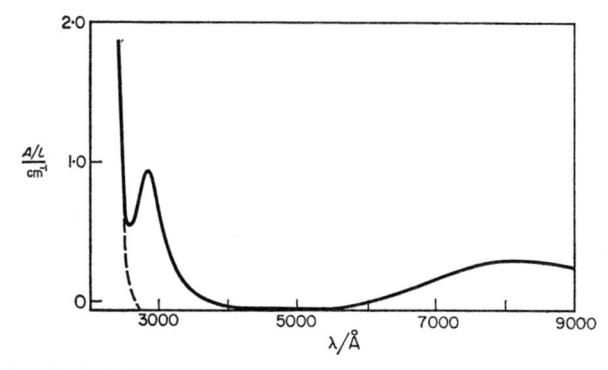

FIG. 2.2. Optical absorption of solution of 0·023M $Cu(ClO_4)_2$ + 0·10M KBr + 0·0044M $HClO_4$ in water. Dotted line shows ultraviolet spectrum when bromide absent. (From R. J. Otter, Ph.D. thesis, University of Reading, 1960.)

region of the spectrum and is of much greater intensity. The electronic transition responsible is one in which an electron is transferred between the central ion and a nearby molecule or ion.[1] For example, formation of $PbCl^+$ produces an intense band with a maximum at 2270 Å and $CuBr^+$ has a band with a maximum at[5] 2810 Å (Fig. 2.2). Absorption of radiation shifts an electron from an orbital essentially located on the halogen to one located on the cation. The intensity of the CuBr+ band is so high that the band appears at concentrations of the complex too low to detect by thermodynamic or conductance measurements. A similarly intense band in the visible is responsible for the blood-red colour of FeSCN^{2+} and higher complexes. These spectroscopic effects bear no direct relation to the nature of the interaction between the ions in the ground state. Even outer-sphere complex formation can produce intense new bands as, for example, with[6] Co(NH$_3$)$_6^{3+}$I$^-$, or substantial changes in bands due to electron transfer between the cation and inner-sphere ligands. If Bjerrum[2] had been able to extend his measurements on the ion Cr(H$_2$O)$_6^{3+}$ to the ultraviolet region of the spectrum he would have found[7] the optical behaviour of the ion was then more dependent on environment. The general principles discussed above also govern anion effects on the absorption spectra of the cations of lanthanides and actinides. The strong absorption bands in the ultraviolet shown by many anions are probably[8] due to electron transfer to molecules of solvent.

For quantitative measurements the spectroscopic method is most powerful when the optical absorption is the specific property of one species involved in a single equilibrium. It becomes less useful for complex ion equilibria with several overlapping stages with each species having their own but similar absorption spectra. Insofar as the molar optical absorption of a species is independent of its concentration, the transmittance T of a solution containing it is given by Beer's Law according to which

$$T = e^{-\kappa l c} \qquad -\ln T = \kappa l c \qquad (2.1.1)$$

where T is the ratio of the intensities of transmitted and incident light, κ is the absorption (extinction) coefficient, l is the length of the light path, and c the concentration of solute. κ is independent of concentration but depends on the nature of the solvent, the temperature and the wavelength of the light. It is common practice to use instead of κ the molar absorptivity (formerly molar extinction coefficient) ϵ defined by

$$\epsilon = \kappa \ln 10 \qquad (2.1.2)$$

and instead of T the absorbance (formerly optical density) A defined by

$$A = -\log T = \epsilon l c \qquad (2.1.3)$$

Beer's Law is only valid for monochromatic light and there are[9] a number of insidious optical causes of apparent deviations from it. It is desirable to make measurements at absorption maxima. The effect on molar absorptivities of small errors in the wavelength setting of the instrument or of shifts in band position due to changes in medium is then least. When, however, both a reactant and a product absorb it is necessary to choose a wavelength for which the difference in their molar absorptivities is large. In a solution containing several absorbing species each of which obeys Beer's Law, $A/l = \Sigma_i \epsilon_i c_i$.

2.2. Simple acid–base equilibria

Changes in the visible spectra of organic acids are the basis of acid–base indicators. A good example of the high precision which can be attained in the quantitative study of an equilibrium even in very

TABLE 2.1
Acidity constant of 2,4-dinitrophenol from spectrophotometric measurements

$10^4 \, c$/mole l^{-1}	0·9245	1·3636	2·3945	3·5827	4·418
α	0·5993	0·5334	0·4410	0·3804	0·3509
$10^5 \, K$/mole l^{-1}	8·143	8·147	8·132	8·143	8·140

dilute solution is provided by the measurements of von Halban and Kortüm[10] of the acidity constant of 2,4-dinitrophenol in water at 25°C. The anion, but not the acid, absorbs light of wavelength 4360 Å. By comparative measurements on solutions of the acid in excess alkali and in water it is possible therefore to determine the concentration of the anion in solutions of the acid in pure water and in various salt solutions. Table 2.1 gives some values of the degree of dissociation α of the acid in solutions of various concentrations c. The acidity constant is given by $K = \alpha^2 c f^2 / (1 - \alpha)$ where f is the mean activity coefficient of the hydrogen and dinitrophenate ions. The activity coefficient of the dinitrophenol molecule is taken as unity. Values of f are calculated from the limiting law as the solutions are so dilute that replacement of $I^{\frac{1}{2}}$ by $I^{\frac{1}{2}}/(1 + I^{\frac{1}{2}})$ only changes f^2 for the most concentrated solution from 0·9714 to 0·9721. The values of K in Table 2.1 are constant within 0·1 per cent, which confirms the exceptionally high quality of the work.

2.3. Use of acid–base indicators

The equilibrium constant for the proton transfer reaction $HX + In \rightleftharpoons X + HIn$ is given by $K = K_{HX}/K_{HIn}$ where K_{HX} and K_{HIn} are the acidity constants of the acids HX and HIn respectively (for simplicity ionic charges are omitted). This is the basis of the familiar indicator method for determining acidity constants ; even if the acid HX and its corresponding base X do not absorb light, the acidity constant K_{HX} can be determined by spectrophotometric measurements of K provided equilibria involving HIn/In can be spectrophotometrically studied. We have

$$K_{HX} = K_{HIn} \frac{[X]}{[HX]} \frac{[HIn]}{[In]} \frac{f_{HIn} f_X}{f_{HX} f_{In}} \qquad (2.3.1)$$

TABLE 2.2

Acidity constant of the hydroxylammonium ion by use of 3,4-dinitrophenol as an indicator

10^2 [HX]/mole l^{-1}	1·417	2·834	4·251	5·668	7·085
A	0·656	0·671	0·678	0·688	0·698
[HIn]/[In]	0·426	0·393	0·378	0·357	0·337
$10^6 K_{HX}$/mole l^{-1} ($B_i = 0$)	1·11	1·14	1·18	1·19	1·17
$10^6 K_{HX}$/mole l^{-1} ($B^i = 0·2$ l mole^{-1})	1·09	1·10	1·11	1·10	1·06

A buffer solution of the acid HX is prepared with a defined [X]/[HX] ratio, a little of the indicator is added and the ratio [HIn]/[In] determined spectrophotometrically. A good example is provided[11] by a determination of the acidity constant of the hydroxylammonium ion using 3,4-dinitrophenol as an indicator. Table 2.2 gives values of A for solutions over a range of [HX] with [X]/[HX] = 0·512 and [NaCl] = [X]. The indicator concentration was $6·8 \times 10^{-5}$M. Both HIn and In absorb light at the wavelength (4000 Å) used. However, provided Beer's Law is obeyed [HIn]/[In] = $(A_{In} - A)/(A - A_{HIn})$ where A_{In} and A_{HIn} are the absorbances for the same concentration of indicator in solutions in which it is completely converted to the base or acid form. 0·01M solutions of NaOH and HCl were used to determine $A_{In} = 0·925$ and $A_{HIn} = 0·024$. The values of the indicator ratio in Table 2.2 are inserted in equation 2.3.1 together with $K_{HIn} = 3·80 \times 10^{-6}$ mole l^{-1}, buffer ratios corrected for a small displacement by the indicator (never greater than 1 per cent), and $\Pi(f)$ calculated from log $\Pi(f) = 2AI^{\frac{1}{2}}/(1 + I^{\frac{1}{2}})$ which is equivalent to setting $\rho = 1$ and $B_i = 0$ in equation 1.3.4. The systematic trend in the values of K_{HX} is removed if $B_i = 0·2$ l mole^{-1} (Table 2.2).

This method is frequently used[12] to find the pK of an organic acid knowing the pK of the buffer acid. To find the pK of a moderately strong acid, when an indicator of comparable pK is difficult to find, it is better[13] to use the indicator to find the concentration of a solution of the acid which has the same hydrogen ion concentration as a solution of a strong acid, e.g. HCl.

2.4. Simple complex ion equilibria

If in an equilibrium M + B $\rightleftarrows$ MB (charges are omitted for simplicity) only the ion M and the complex MB absorb light, and their molar absorptivities are constant, then

$$A/l = \epsilon_{\text{M}}[\text{M}] + \epsilon_{\text{MB}}[\text{MB}] = \epsilon_{\text{M}}(m-x) + \epsilon_{\text{MB}}x \tag{2.4.1}$$

Further, if the ionic activity coefficients are maintained constant by use of a swamping ionic medium, then

$$Q = [\text{MB}]/[\text{M}][\text{B}] = x/(m-x)(b-x) \tag{2.4.2}$$

ϵ_{M} can usually be determined by separate experiments in the absence of B, so the pair of equations have to be solved for Q and ϵ_{MB}. The combined equations have been written in various forms, unnecessarily known by proper names. One common form for the special case $\epsilon_{\text{M}} = 0$, obtained by substituting $A/l\epsilon_{\text{MB}} = x$ in $x = m - x/Q(b-x)$ is,

$$A/l = \epsilon_{\text{MB}}m - A/lQ(b-x) \tag{2.4.3}$$

which is solved iteratively by plotting A against $A/(b-x)$. Alternatively substituting $A/l\epsilon_{\text{MB}} = x$ in $x = mbQ - mxQ - bxQ + x^2Q$ one readily obtains

$$\frac{lm}{A} = \frac{1}{\epsilon_{\text{MB}}bQ} + \frac{m}{b\epsilon_{\text{MB}}} + \frac{1}{\epsilon_{\text{MB}}} - \frac{A}{l\epsilon_{\text{MB}}^2 b} \tag{2.4.4}$$

With a complex of high extinction coefficient such as FeSCN^{2+} and with b constant and $m \gg b$, the last two terms can be neglected and ϵ_{MB} and Q are obtained from a linear plot of lm/A against m. From such a plot Lister and Rivington[14] obtain for FeSCN^{2+} in 1·2M NaClO$_4$ + 0·2M HClO$_4$ ($b = 1·5 \times 10^{-4}$M, $m = 0·002$ to $0·02$M) values of $Q = 176$ l mole^{-1}, $\epsilon_{\text{FeSCN}^{2+}} = 4728$ cm^{-1} l mole^{-1} at $\lambda = 4500$ Å. Such molar absorptivity is typical of values for electron-transfer bands. Note that $\epsilon = 5 \times 10^3$cm^{-1} l mole^{-1} corresponds to an absorption cross-section of 5×10^3cm^{2}m.mole$^{-1} = 0·08$ Å^2 molecule^{-1}. In some cases, for example[5] CuBr$^+$, only the first term on the right-hand side of equation 2.4.4 is appreciable and only the product $\epsilon_{MB}Q$ can be determined. Equations such as 2.4.3 and 2.4.4 are, of course, invalid if more than one complex is formed. With ferric thiocyanate, the formation of Fe(SCN)$_2^+$ and mixed complexes of FeSCN^{2+} with halide and sulphate ions has been studied[14] spectrophotometrically, but satisfactory analysis of the results is difficult (see 2.5).

In a lengthy series of careful measurements on the equilibrium Fe^{3+} + OH$^-$ $\rightleftarrows$ FeOH^{2+} in perchlorate media ($\lambda = 2900\text{–}3300$ Å), Richards and Sykes[15] have determined both Q and the extinction coefficients at several ionic strengths and fitted the Q values to an expression for the thermodynamic constant K with $K = 6\cdot6 \times 10^{11}$ l mole^{-1} at 25°C. Changing the perchlorate ion concentration whilst keeping the ionic strength the same, which can be done by replacing the univalent sodium cation by a higher valent one (Ba^{2+} or La^{3+}) has a specific effect on Q. Richards and Sykes ascribe this to outer-sphere complex formation between Fe^{3+} and ClO$_4^-$, with no effect on the extinction coefficient of Fe^{3+}.

In suitable cases, thermodynamic equilibrium constants can be obtained from measurements in the absence of other non-reacting electrolytes. The dimerization of the hydrogen chromate ion in aqueous solution, 2HCrO$_4^-$ $\rightleftarrows$ Cr$_2$O$_7^{2-}$ + H$_2$O is a good example[9] where although neither extinction coefficient can be separately determined, it is possible to determine both these and the equilibrium constant from precise measurements on $0\cdot3 \times 10^{-4}$ to 7×10^{-4}M solutions of potassium dichromate (in the presence of a trace of acid to repress formation of CrO$_4^{2-}$). For the equilibrium 2M $\rightleftarrows$ D

$$K = 4\alpha^2 c f_M^2/(1-\alpha) f_D \qquad (2.4.5)$$

$$A/lc = (1-\alpha)\epsilon_D + 2\alpha\epsilon_M = \epsilon_D - (\epsilon_D - 2\epsilon_M)\alpha \qquad (2.4.6)$$

and as the ionic strength never exceeded $0\cdot022$M it can also be assumed with reasonable confidence that

$$\log(f_M^2/f_D) = 2AI^{\frac{1}{2}}/(1+I^{\frac{1}{2}}) \qquad (2.4.7)$$

The value of K is found which leads to a linear plot of A/lc against α. Table 2.3 shows the fit of the results for one set of measurements. For each set K could be located within about 2 per cent. The average value from several sets was $0\cdot030$ mole l^{-1} which is in quite good agreement with one of $0\cdot028$ mole l^{-1} obtained independently[16] by extrapolating values of Q determined at several ionic strengths in perchlorate media.

The results of applying this mode of analysis to measurements on copper sulphate solutions are instructive.[17] As the concentration increases there is a small change in the absorption of the copper ion in the visible and a pronounced one in the ultraviolet. A sharply rising absorption below 2500 Å shifts to longer wavelengths in the presence of increasing concentrations of sulphate ions. For example, the molar absorptivity at 2400 Å rises from about 83 cm^{2}m.mole^{-1} in a 1×10^{-3}M solution to 140 cm^{2}m.mole^{-1} in a 25×10^{-3}M solution.

TABLE 2.3

Extinction coefficients and thermodynamic equilibrium constant for $2HCrO_4^- \rightleftharpoons Cr_2O_7^{2-} + H_2O$
(ϵ(calc) from $K = 0.0285$ mole l^{-1}, $\epsilon_D = 1742$ cm^{2}m.mole^{-1}, $\epsilon_M = 255$ cm^{2}m.mole^{-1})

$10^4 c$/mole l^{-1}	71.45	19.45	11.01$_4$	6.085	3.241
ϵ/cm^{2}m.mole^{-1}	1045.4	769.4	679.4	614.1	571.8
ϵ(calc)/cm^{2}m.mole^{-1}	1045.6	768.3	680.2	614.7	569.9

This is in contrast with the behaviour of nickel and cobalt sulphates, and is probably because two of the six water molecules of the hydrated copper ion are easily displaced. If we postulate an equilibrium $Cu^{2+} + SO_4^{2-} \rightleftharpoons CuSO_4$, the absorbance of the solution can be used to measure the distribution of cupric ions between free and associated classes. If the associated class contains both inner and outer-sphere complexes, it may well be only in the former sub-class that cupric ions are optically different from free cupric ions. Nevertheless provided the relative populations of the inner and outer-sphere sub-classes are independent of concentration the molar absorptivity of associated ions will be an average over the two sub-classes. For the equilibrium $CuSO_4 \rightleftharpoons Cu^{2+} + SO_4^{2-}$ if one writes

$$K = 4\alpha^2 c f^2/(1-\alpha)c \qquad (2.4.8)$$

and
$$A/lc = (1-\alpha)\epsilon_{CuSO_4} + \alpha\epsilon_{Cu} = \epsilon_{CuSO_4} - (\epsilon_{CuSO_4} - \epsilon_{Cu})\alpha \qquad (2.4.9)$$

$$-\log f^2 = 8AI^{\frac{1}{2}}/(1+\rho I^{\frac{1}{2}}) \qquad (2.4.10)$$

it is found that for concentrations up to 0.025M the results can be equally well fitted (better than 0.5 per cent) by several sets of values of the parameters (Table 2.4). The probable reason for this flexibility is that both short and long-range interactions of the cupric and sulphate ions are of an essentially ionic nature. There is an appreciable population of pairs of cupric and sulphate ions which can be approximately treated as belonging to either the free or the associated class. A plot of goodness of fit against values of K passes through only a flat maximum. The dissociation constants have well-defined significance only in relation to a specified expression for the activity coefficient of the free ions. An apparent restriction of the range of parameters could be achieved by importing a molar absorptivity for the complex ion determined from measurements of the equilibrium in a swamping ionic medium. In (say) 2M sodium perchlorate copper and sulphate ions will behave as associated only at very close distances of approach; at greater distances their interactions with each other will be negligible compared with interactions with ions of the swamping electrolyte.

TABLE 2.4
Extinction coefficients and thermodynamic dissociation constant for $CuSO_4 \rightleftharpoons Cu^{2+} + SO_4^{2-}$

$10^3 K$/mole l^{-1}	8·0	4·0	3·5
d/Å	4·3	10·0	14
ϵ_{CuSO_4}/cm^{2}m. mole^{-1}	353·9	222·2	205·4
ϵ_{Cu}/cm^{2}m. mole^{-1}	62·2	62·3	61·9

Other examples of complexes the stabilities of which have been studied spectrophotometrically are[18] $CoSCN^{2+}$, $CoCl^+$, $CoCl_4^{2-}$, CoS_2O_3, $CuOAc^+$, MgS_2O_3, $CdNO_2^+$, $PbCl^+$, $PbBr^+$, $TlOH$, UO_2SO_4, [19]NpO_2SO_4,[19] $NpSO_4^{2+}$, $CeSO_4^+$, $FeSO_4^+$, $Co(NH_3)_5SO_4^+$. Complexes of the malonate ion with alkaline-earth cations have been studied[20] by measuring with an acid–base indicator the shift in pH of a malonate buffer on addition of the cations. Species of the outer-sphere type which have been studied spectrophotometrically by ultraviolet measurements are[18] $Co(NH_3)_6^{3+}SO_4^{2-}$, $Co(NH_3)_5H_2O^{3+}SO_4^{2-}$, $Mg^{2+}Fe(CN)_6^{4-}$, $Ba^{2+}Fe(CN)_6^{4-}$, $La^{3+}Fe(CN)_6^{4-}$, $Tl^+Fe(CN)^{4-}$, $K^+Fe(CN)_6^{4-}$. When an optical effect is small and the complex is of low stability, the physical significance of parameters obtained by fitting results to equations such as 2.4.3 should be treated with scepticism. It is necessary to make measurements over a wide range of ionic composition and even if the formal ionic strength is kept constant, it is unlikely either that the effects of the change in composition on the optical behaviour of the system are adequately described by a simple equilibrium or that the ionic activity coefficients remain constant. Two cases which have been carefully studied and discussed are those of[21] $FeCl^{2+}$ and[22] PuO_2Cl^+. There is an apparent dependence of the values of Q on wavelength. Ideally, measurements on any system should be repeated at several wavelengths. Only a few systems have been studied in non-aqueous solvents. Among these are the complexes of the alkaline-earth cations with picrates in methanol[23] and triphenylmethyl chloride in sulphur dioxide.[24] The latter is an interesting system, for it seems there are three states which can be represented by $Ph_3C-Cl \rightleftarrows Ph_3C^+Cl^- \rightleftarrows Ph_3C^+ + Cl^-$. A self-consistent interpretation of spectrophotometric and conductance measurements can be given by assuming that the spectrophotometric change occurs on ionization; it is only when the pair of ions formed dissociate that they can make a conductance contribution.

2.5. Overlapping complex ion equilibria

The spectrophotometric method is not well suited to the determination of the successive constants of overlapping equilibria, for each species has an unknown extinction coefficient as well as its stability constant. Various ways of solving the complicated simultaneous equations have been suggested, but it is not often that reliable conclusions emerge. It is, however, possible[25] to use the spectrophotometric method to determine not equilibrium constants directly but free ligand concentrations in the solutions. Consider a set of overlapping equilibria $M + nB \rightleftarrows MB_n$ in which the ligand does not absorb light (e.g. $M = Cu^{2+}$, $B = NH_3$). The optical absorbance is given by

$$\frac{A}{lm} = \frac{\sum_{n=0}^{N} \epsilon_n [MB_n]}{\sum_{n=0}^{N} [MB_n]} = \frac{\epsilon_0 [M] + \sum_{n=1}^{N} \epsilon_n Q_n [M][B]^n}{[M] + \sum_{n=1}^{N} Q_n [M][B]^n} = F_1([B])$$

(2.5.1)

where m is the stoichiometric concentration of M and ϵ_n and Q_n are the molar absorptivities and stability constants for the species MB_n. A/lm is a function of the free ligand concentration alone. Therefore a pair of solutions with different values of m and b, the stoichiometric concentration of ligand, but identical values of A/lm, have identical concentrations of free ligand. Such solutions are known as " corresponding solutions ". They also have identical values of what is known as the ligand number $\bar{n}$ defined by

$$\bar{n} = \frac{b - [B]}{m} = \frac{\sum\limits_{n=1}^{N} nQ_n[M][B]^n}{[M] + \sum\limits_{n=1}^{N} Q_n[M][B]^n} = F_2([B]) \qquad (2.5.2)$$

and the equation $(b' - [B])/m' = (b'' - [B])/m''$ can be solved for $[B]$. The results are then analysed in the same way as those obtained by, for example, e.m.f. measurements.

What is often done by those interested in the absorption spectra of the species themselves is to combine spectrophotometric measurements with stability constants determined by other methods to obtain the absorption bands of each species. This has been done, for example,[26] for each of the species from $Ni(NH_3)^{2+}$ to $Ni(NH_3)_6^{2+}$.

2.6. Raman spectroscopy

The study of infrared absorption spectra in aqueous solution is in its infancy because of the difficulties introduced by the intense absorption of the solvent. Precise photoelectric measurements of Raman spectra have been made particularly by Young and his co-workers.[27] A careful study has been made[28] for a series of nitrates of changes with concentration of the frequency, the half-width and the molar integrated intensity (area/concentration) of the 1048 cm^{-1} nitrate ion line. The molar integrated intensity is less dependent on the ionic background than the half-width or the molar intensity at a particular wavelength, but nevertheless there are changes of a few per cent over the concentration range up to 10M which are specifically dependent on the cation. The very small change shown by ammonium nitrate is ascribed to the similarity between the ammonium ion and the water molecule. Similarly the integrated intensity of the 980 cm^{-1} line of the sulphate ion in aqueous ammonium sulphate solutions shows[27] almost perfect proportionality with molality up to 10M. As an ammonium ion and a water molecule are so similar in their effect it is reasonable to assume that the same is true of a hydroxonium ion and a water molecule. Well separated characteristic lines for strong acid molecules such as H_2SO_4 and HNO_3 and for their anions, together with the assumption of proportionality between the integrated intensity of a line and species concentration, has made possible[27] the determination of degrees of dissociation of these acids in solutions of concentrations up to those of the anhydrous acids. Figure 2.3 shows the degree of dissociation of nitric acid in water as a function of concentration determined by Young's school, and Fig. 2.4 the concentration of the molecular species in aqueous sulphuric acid. The intensity of the 1049 cm^{-1} nitrate line was used in the study of nitric acid, whilst in the study of sulphuric acid the 980 cm^{-1} SO_4^{2-} and 1040 cm^{-1} HSO_4^- lines gave estimates of the concentrations of the respective ions and by difference the concentration of H_2SO_4 which was accurately proportional to the observed intensity of its 910 cm^{-1} line. Of course

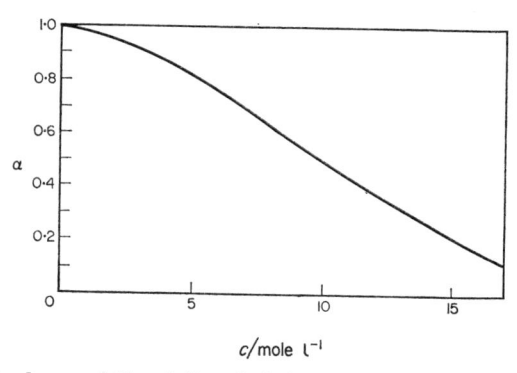

Fig. 2.3. The degree of dissociation of nitric acid in water at 25°C determined by Raman spectroscopy.

for such concentrated solutions it is out of the question to obtain the thermodynamic dissociation constant from α values and theoretical estimates of the activity coefficients. However, if the stoichiometric ionic activity $a_s = m_s^2 \gamma_s^2$ is known from independent thermodynamic measurements, as is the case with nitric acid, the dissociation constant can be calculated from

$$K = \frac{(\alpha m \gamma_i)^2}{(1-\alpha)m\gamma_u} = \frac{(m\gamma_s)^2}{(1-\alpha)m\gamma_u} = \frac{m\gamma_s^2}{(1-\alpha)\gamma_u} = \frac{K'}{\gamma_u} \quad (2.6.1)$$

γ_u is unknown but if it is assumed that $\log \gamma_u = B_u(1-\alpha)m + B_i \alpha m$ (cf. the linear term of equation 1.3.6) which can be written $\log \gamma_u = B_i \{r + (1-r)\alpha\}m$ with $r = B_u/B_i$, K can be obtained from a plot of $\log K'$ against m. The value of r is selected to give the best linear plot. By putting $r = 0.7$, McKay[29] calculates from Young and Kravetz's results from 2 to 8M that at 25°C, $K = 23.5 \pm 0.5$ mole kg^{-1} with $B_i = 0.048$ kg mole^{-1}.

Quantitative Raman work so far done has been almost exclusively restricted to the strong acids. Even here the overlap of lines from the undissociated acid and the anion may be so marked, e.g. iodic acid,[30] that measurement of the integrated intensity of a line associated with a single species is impossible. One has then to study the intensity of a line at a particular wavelength as a function of concentration, just as in ultraviolet or visible spectrophotometry the variation with concentration of optical density at a particular wavelength is studied. Raman spectroscopy is also limited by the low intensity of many lines especially

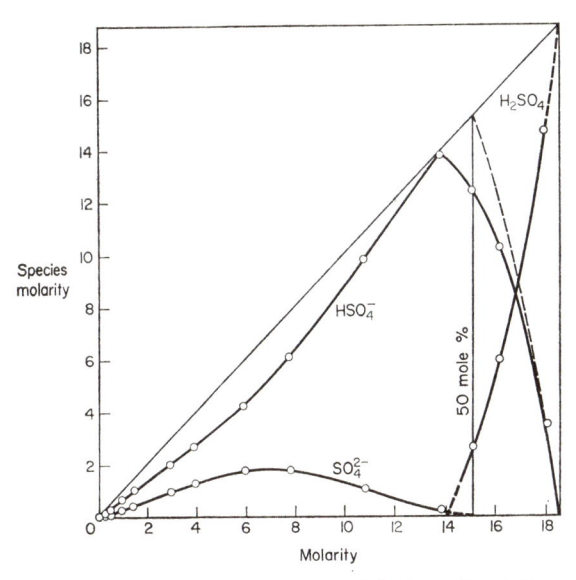

Fig. 2.4. Concentration of species in aqueous sulphuric acid solutions at 25°C. The middle broken curve gives the HSO_4^- concentration if the reaction $H_2O + H_2SO_4 \rightleftarrows H_3O^+ + HSO_4^-$ were complete. (From reference 27.)

in the case of the more polar bonds, for the change in polarizability with vibration, a necessary condition for the occurrence of a Raman line, will then be very small.

The independence of environment of the Raman spectrum of a species may sometimes mean that interactions most conveniently treated as ionic equilibria in other contexts are undetected. In 2.4, ultraviolet absorption measurements on copper sulphate solutions were interpreted by invoking the incomplete dissociation of the salt. Later in the book the same model will be used to interpret conductance and thermodynamic measurements on this and other sulphates of bivalent metals. Yet the Raman spectra of solutions of these sulphates[31] show no significant changes in the characteristic spectrum of the free sulphate ion. This is also true of the sulphates of Ga^{3+} and Al^{3+}, and it is in the case of In^{3+} alone that new Raman lines characteristic of a complex appear. The base $(CH_3)_2TlOH$ is according to kinetic[32] and conductimetric measurements only 50 per cent dissociated in 0·2M solution, and yet the lines of the Raman spectrum of the $(CH_3)_2Tl^+$ ion in a 2M solution are only slightly shifted compared with those in a nitrate or perchlorate solution.[33] Their integrated intensities may well be identical. No new line is detectable which could be associated with a thallium–oxygen vibration. It seems that unless short-range ionic interactions, however strong, are of a pronounced covalent nature, they will only show themselves in the Raman spectrum by slight shifts in the positions and shapes of the lines of the interacting ions. New lines should appear in the infrared spectrum, for an ionic bond will have a large change of dipole moment associated with its vibration. So far, however, quantitative studies by infrared spectroscopy of ionic equilibria are limited to stable complexes such as the cyano-complexes of nickel in aqueous solution.[34]

2.7. Nuclear magnetic resonance spectroscopy

The n.m.r. frequency of a proton or other nucleus depends on the molecule in which it is, for the magnetic field at a particular nucleus contains a contribution from the nuclei and electrons in its immediate vicinity as well as from the applied field. This means that the magnetic resonance frequency of a proton is different in (say) H_2O, H_3O^+ and HNO_3; the differences only amount to a few millionths of the resonance frequency but can be measured with high accuracy. The quantity usually measured is the shift of applied field required for resonance at a fixed frequency relative to that for a reference substance, e.g. H_2O. If the exchange rate of the nucleus between two states is fast compared with the separation of the frequencies corresponding to the two molecules, only a single line is observed but its position depends on the relative numbers of nuclei in the two states. Proton exchanges give rise to two lines only if the half time of exchange is greater than about 10^{-3} sec.

The use of the n.m.r. method for studying the dissociation of mineral acids depends on the insensitivity of the chemical shift of the proton to changes of environment outside the molecule in which it is situated. If measurements are made on a partially ionized acid $HA(HA + H_2O \rightleftarrows H_3O^+ + A^-)$, then if environmental effects on the shift for H_2O, H_3O^+

and HA can be neglected, the measured shift δ against water as a standard will be given by

$$\delta/p = \delta_{H_3O}\alpha + \tfrac{1}{3}\delta_{HA}(1-\alpha) \qquad (2.7.1)$$

α is the degree of dissociation of HA; p is the fraction of hydrogen nuclei in the solution present in H_3O^+ if protolysis were complete, and δ_{H_3O} and δ_{HA} are the chemical shifts for a hydrogen nucleus in H_3O^+ and HA respectively. Small corrections for the differences in the bulk magnetic susceptibilities of the acid solutions and the reference solutions are necessary. For hydrochloric and nitric acid solutions,[35] δ is the same within experimental error and a nearly linear function of p up to a concentration of several molar. Over this range in both cases α is effectively unity and a value for δ_{H_3O} can be obtained. For nitric acid the δ–p plot becomes non-linear over the same concentration range that Raman measurements indicate incomplete dissociation. If a value of α close to zero for a 15·95M solution is accepted from the Raman measurements δ_{HA} can be obtained. The α values calculated for other concentrations from equation 2.7.1 are then in satisfactory agreement with the results of the Raman method. The values of α have also received general confirmation[36] by study of the ^{14}N shifts.

Results[37] for sulphuric acid solutions are also consistent with conclusions from Raman measurements, and another strong acid which has been studied[35] by the n.m.r. method is perchloric. The position is not in this case entirely satisfactory. The value of δ_{H_3O} obtained from measurements on solutions of moderate concentration is about 20 per cent lower than the value obtained from measurements on hydrochloric and nitric acids. It seems that the effects of the perchlorate ion on δ_{H_2O} or δ_{H_3O} are not negligible.[35a]

Interactions too weak to be detected by the Raman method can cause shifts in n.m.r. frequencies. For example, the shift in the ^{205}Tl frequency when increasing concentrations of potassium hydroxide are added to a solution of thallous hydroxide is consistent[38] with kinetic and ultraviolet spectrophotometric estimates of the degree of formation of undissociated TlOH. Even in concentrated solutions there is no sign of a Raman line associated with this species. Concentration dependence of cation and halide ion shifts in concentrated alkali halide solutions occurs with the larger ions,[39] e.g. caesium bromide. This suggests that ion–ion interaction is more pronounced in these cases, which is consistent with other evidence (see 6.3 and 10.5).

REFERENCES

1. ORGEL, L. E., *An Introduction to Transition-Metal Chemistry* : *Ligand Field Theory*, Methuen, London, 1960, p. 86.
2. BJERRUM, N., *Z. anorg. Chem.*, 1909, **63**, 140.
3. HOLMES, O. G. and MCCLURE, D. S., *J. chem. Phys.*, 1957, **26**, 1686.
4. SMITHSON, J. P. and WILLIAMS, R. J. P., *J. chem. Soc.*, 1958, 457.
5. HOPE, D. A. L., OTTER, R. J. and PRUE, J. E., *J. chem. Soc.*, 1960, 5226 ; OTTER, R. J., Ph.D. thesis, University of Reading, 1960.
6. LINHARD, M., *Z. Elektrochem.*, 1944, **50**, 224.
7. GATES, H. S. and KING, E. L., *J. Amer. chem. Soc.*, 1958, **80**, 5011.
8. ORGEL, L. E., *Quart. Rev. chem. Soc., Lond.*, 1954, **8**, 422.
9. DAVIES, W. G. and PRUE, J. E., *Trans. Faraday Soc.*, 1955, **51**, 1045.
10. VON HALBAN, H. and KORTÜM, G., *Z. phys. Chem. A.*, 1924, **170**, 351.
11. ROBINSON, R. A. and BOWER, V. E., *J. phys. Chem.* 1961, **65**, 1279.
12. ROBINSON, R. A., *The Structure of Electrolytic Solutions*, Ed. HAMER, W. J., John Wiley, New York, 1959, p. 253.
13. VON HALBAN, H. and BRÜLL, J., *Helv. chim. acta*, 1944, **27**, 1719.
14. LISTER, M. W. and RIVINGTON, D. E., *Canad. J. Chem.*, 1956, **33**, 1572, 1591, 1603.
15. RICHARDS, D. H. and SYKES, K. W., *J. chem. Soc.*, 1960, 3626.
16. TONG, J. Y.-P. and KING, E. L., *J. Amer. chem. Soc.*, 1953, **75**, 6180.
17. DAVIES, W. G., OTTER, R. J. and PRUE, J. E., *Disc. Faraday Soc.*, 1957, **24**, 103.
18. PRUE, J. E., *Rep. Progr. Chem.*, 1958, **55**, 14 ; 1960, **57**, 79 ; COVINGTON, A. K. and PRUE, J. E., *Rep. Progr. Chem.*, 1963, **60**, 55.
19. TAYLOR, B. L. and SYKES, K. W., *Proc. 7th Int. Conf. Co-ordination Chemistry*, Stockholm, 1962, p. 31.
20. STOCK, D. I. and DAVIES, C. W., *J. chem. Soc.*, 1949, 1371.
21. WOODS, SR. M. J. M., GALLAGHER, P. K. and KING, E. L., *Inorg. Chem.*, 1962, **1**, 55.
22. NEWTON, T. W. and BAKER, F. B., *J. phys. Chem.*, 1957, **61**, 934.
23. KORTÜM, G. and ANDRUSSOW, K., *Z. phys. Chem. (Frankfurt)*, 1960, **25**, 321.
24. POCKER, Y., *Proc. chem. Soc.*, 1959, 386.
25. BJERRUM, J., *Kgl. danske Videnskab. Selskab, Mat.-fys. Medd.*, 1944, **21**, No. 4.
26. BJERRUM, J., *Metal Ammine Formation in Aqueous Solution*, P. Haase, Copenhagen, 1941, p. 190.
27. YOUNG, T. F., MARANVILLE, L. F. and SMITH, H. M., *The Structure of Electrolytic Solutions*, Ed. HAMER, W. J., John Wiley, New York, 1959, p. 35.
28. VOLLMER, P. M., *J. chem. Phys.*, 1963, **39**, 2236.
29. MCKAY, H. A. C., *Trans. Faraday Soc.*, 1956, **52**, 1568.
30. HOOD, G. C., JONES, A. C. and REILLY, C. A., *J. phys. Chem.*, 1959, **63**, 101.
31. HESTER, R. E., PLANE, R. A. and WALRAFEN, G. E., *J. chem. Phys.*, 1963, **38**, 249.
32. LAWRENCE, J. K. and PRUE, J. E., "Int. Conf. Co-ordination Chem.," *Chem. Soc. Special Publ.*, 1959, No. 13, p. 186.
33. GOGGIN, P. L. and WOODWARD, L. E., *Trans. Faraday Soc.*, 1960, **56**, 1591.
34. MCCULLOUGH, R. L., JONES, L. H. and PENNEMAN, R. A., *J. inorg. nuclear Chem.*, 1960, **13**, 286.
35. HOOD, G. C., REDLICH, O. and REILLY, C. A., *J. chem. Phys.*, 1954, **22**, 2067 ; HOOD, G. C. and REILLY, C. A., *J. chem. Phys.*, 1960, **32**, 127.
35a. AKITT, J. W., COVINGTON, A. K., FREEMAN, J. G. and LILLEY, T. H., *Chem. Comm., Lond.*, 1965, 349.
36. MASUDA, Y. and KANDA, T., *J. Phys. Soc. Japan*, 1953, **8**, 432.
37. HOOD, G. C. and REILLY, C. A., *J. chem. Phys.*, 1957, **27**, 1126.
38. FREEMAN, R., GASSER, R. P. H., RICHARDS, R. E. and WHEELER, D. H., *Mol. Phys.*, 1959, **2**, 75.
39. CRAIG, R. A., *Rep. Progr. Chem.*, 1962, **59**, 63.

CHAPTER 3

CONDUCTANCE

3.1. Introduction

Electrolytic conductance is the most direct evidence for the existence of ions in solution, and in solutions of a single electrolyte its variation the most obvious way of studying ionic equilibria. Measurements of high precision can be made on very dilute solutions in a wide range of solvents. The degree of (single-stage) dissociation into ions of a neutral species at a concentration c is given by the simple relation $\alpha = \Lambda/\Lambda_i$ where Λ is the measured molar conductance of the solution and Λ_i is the hypothetical molar conductance the electrolyte would have if it were completely dissociated into ions at a concentration αc. Arrhenius' original suggestion[1] was that $\alpha = \Lambda/\Lambda_0$, that is to equate Λ_i to Λ_0, the molar conductance of the electrolyte at infinite dilution. This, however, neglects the effect of ionic environment on the mobility of ions due to long-range coulombic forces. It is this kind of interaction which causes the decrease with increasing concentration of the molar conductance values for sodium acetate in Table 1.1. As the table shows, any attempt to interpret the decrease as due to an electrolytic dissociation equilibrium fails. A subtle compensation of errors is partly responsible for the constancy of the dissociation constant for acetic acid given in the same table. The classical interpretation underestimates α and overestimates $f_\pm$ (which is set equal to unity) progressively more as the concentration increases. It happens that these two effects roughly cancel in their influence on $\alpha f_\pm$ in water, but this would not be so in other solvents.

3.2. Weak electrolytes

One of the first precise values for the thermodynamic acidity constant of acetic acid was obtained by MacInnes and Shedlovsky.[2] They calculated α values from conductance measurements on acetic acid using values of Λ_i obtained by assuming that ionic mobilities are additive at finite concentrations. If this is valid and $\alpha = 1$ for the electrolytes HCl, NaCl and NaOAc then

$$\Lambda_i = \Lambda_{HCl} - \Lambda_{NaCl} + \Lambda_{NaOAc} \tag{3.2.1}$$

Some of the values of α obtained by MacInnes and Shedlovsky are given in the third column of Table 3.1 (three successive approximations are necessary in computing Λ_i at an ionic concentration of αc). Even though the ionic concentrations are very low throughout with

TABLE 3.1
Acidity constant of acetic acid in water at 25°C from conductance data
($\Lambda_0 = 390\cdot71$ int. Ω^{-1} cm^2 mole^{-1})

$10^3 c$ mole l^{-1}	Λ int. Ω^{-1} cm^2 mole^{-1}	α McI. and S.	α' L.L.	$10^5 K_a$ mole l^{-1}
0·11135	127·71	0·3278	0·3276	1·748
0·21844	96·466	0·2477	0·2476	1·750
1·02831	48·133	0·1238	0·12374	1·750
2·41400	32·208	0·08290	0·08289	1·749
5·91153	20·956	0·05401	0·05401	1·748
20·000	11·563	0·02987	0·02988	1·734
50·000	7·356	0·01905	0·01905	1·719
100·000	5·200	0·013496	0·013501	1·694
200·000	3·650	0·009495	0·009502	1·645

$\alpha c < 0\cdot002$M, it is possible that the conductances of HCl, NaCl and NaOAc are affected by specific interactions which are absent when the only ions in solution are hydrogen and acetate ions. It is therefore interesting and, with a view to general application, important to be able to calculate Λ_i from some theoretical equation. Equation 3.2.1 is replaced by the analogous formula for the Λ_0's, so that Λ_0 of the incompletely dissociated acid is assumed to be known. If within the concentration range studied Λ_i is given by Onsager's limiting law,[3]

$$\frac{\Lambda_i}{\Lambda_0} = 1 - \left(\theta + \frac{\sigma}{\Lambda_0}\right)(\alpha c)^{\frac{1}{2}} \tag{3.2.2}$$

where θ and σ are coefficients of the theory which depend on the nature of the solvent. For a symmetrical electrolyte of charge number z, θ is replaced by $z^2\theta$, σ by $z^2\sigma$ and αc by I, the ionic strength. Combination of the pair of equations $\Lambda/\Lambda_i = \alpha$ and 3.2.2 gives

$$\frac{\Lambda}{\alpha\Lambda_0} = 1 - \left(\theta + \frac{\sigma}{\Lambda_0}\right)(\alpha c)^{\frac{1}{2}} \tag{3.2.3}$$

which is a cubic equation for $\alpha^{\frac{1}{2}}$. It can be solved by successive approximations or by use of a table published[4] by Fuoss to expedite the numerical solution. Introducing the abbreviations

$$F = \Lambda/\alpha\Lambda_0 \tag{3.2.4}$$

$$Z = \left(\theta + \frac{\sigma}{\Lambda_0}\right)\frac{\Lambda^{\frac{1}{2}}}{\Lambda_0}c^{\frac{1}{2}} \tag{3.2.5}$$

equation 3.2.3 may be rewritten

$$F = 1 - ZF^{-\frac{1}{2}} \tag{3.2.6}$$

which determines F and so also α for a given value of the experimental quantity Z. Values of F as a function of Z have been tabulated for values of Z up to 0·209. The values of α' in Table 3.1 are obtained from equation 3.2.3.

The limiting law is valid only when the ionic concentration is very small. Accurate measurements on sodium and potassium chloride solutions show that the law is obeyed in aqueous solution to within ±0·01 per cent below 0·001M. Shedlovsky[5] made the empirical observation that for some strong uni-univalent electrolytes in water the equation

$$\frac{\Lambda}{\Lambda_0} = 1 - \frac{\Lambda}{\Lambda_0}\left(\theta + \frac{\sigma}{\Lambda_0}\right)c^{\frac{1}{2}} \tag{3.2.7}$$

is valid up to about 0·01M. It only differs from the limiting law by a trivial term proportional to c and has the advantage that for an incompletely dissociated electrolyte one obtains

$$\frac{\Lambda}{\alpha\Lambda_0}\left\{1 + \left(\theta + \frac{\sigma}{\Lambda_0}\right)\alpha^{\frac{1}{2}}c^{\frac{1}{2}}\right\} = 1 \tag{3.2.8}$$

which is quadratic in $\alpha^{\frac{1}{2}}$ and more readily soluble than equation 3.2.3. Using the abbreviations 3.2.4 and 3.2.5, equation 3.2.8 can be rewritten

$$F = 1 - ZF^{\frac{1}{2}} \tag{3.2.9}$$

It is even simpler to use instead of equations 3.2.6 or 3.2.9

$$F = 1 - Z \tag{3.2.10}$$

which is linear in α and directly obtained by substituting the first approximation $\alpha^{\frac{1}{2}} = (\Lambda/\Lambda_0)^{\frac{1}{2}}$ in equation 3.2.3.

Of greater importance than empirical extensions of the limiting law is its theoretical extension by removal of the restriction $\kappa d \ll 1$ where

κ^{-1} is the radius of the ion atmosphere and d the mean ionic diameter; κd is equal to $\rho I^{\frac{1}{2}}$ used earlier. This leads to

$$\frac{\Lambda}{\Lambda_0} = 1 - \frac{\theta c^{\frac{1}{2}}}{(1 + \rho c^{\frac{1}{2}})(1 + 2^{-\frac{1}{2}}\rho c^{\frac{1}{2}})} - \frac{\sigma c^{\frac{1}{2}}}{\Lambda_0(1 + \rho c^{\frac{1}{2}})} \qquad (3.2.11)$$

as a first approximation[6,7,8]. It will be convenient to refer to this as the Leist equation. Robinson and Stokes[9] proposed the simplified equation

$$\frac{\Lambda}{\Lambda_0} = 1 - \left(\theta + \frac{\sigma}{\Lambda_0}\right)\frac{c^{\frac{1}{2}}}{1 + \rho c^{\frac{1}{2}}} \qquad (3.2.12)$$

which gives a useful ($\pm 0\cdot 1$ per cent) but not exact fit of data for strong uni-univalent electrolytes in water up to about $0\cdot 1\text{M}$ with reasonable values of ρ. This equation is less satisfactory for non-aqueous solutions. Unfortunately there is disagreement at present about the importance of higher terms in the conductance equation. Equations have been proposed by Pitts[7] and in several versions by Fuoss and Onsager.[8] These equations are used with tabulated functions of κd and have been tested[10,11,12] with data for completely dissociated uni-univalent electrolytes. The equations differ in their predictions to an extent which depends on the solvent properties, Λ_0 and z for the electrolyte, and the value of d. A decisive choice between them is so far not possible. A number of workers have used the Fuoss–Onsager equation in a simplified form for the analysis of data for incompletely dissociated electrolytes. The simplified equation is

$$\Lambda = \Lambda_0 - (\theta \Lambda_0 + \sigma)c^{\frac{1}{2}} + Ec \log c + Jc \qquad (3.2.13)$$

where E depends on Λ_0 and the solvent properties and J on these and also on d.

If $d = 4$ Å, equation 3.2.11 gives $\alpha = 0\cdot 009488$ for the result in the last row of Table 3.1. This is for the most concentrated solution with $\kappa d = 0\cdot 06$. The ionic concentration for the results in Table 3.1 does not exceed $0\cdot 002\text{M}$ and it is immaterial whether the limiting law or an extension of it is used for calculating α. The various estimates never differ by as much as $0\cdot 15$ per cent. A value of α having been obtained at each concentration, values of K_a are calculated from

$$K_a = \frac{\alpha^2 cf^2}{1 - \alpha} \qquad (3.2.14)$$

$$-\log f = \frac{A(\alpha c)^{\frac{1}{2}}}{1 + \rho(\alpha c)^{\frac{1}{2}}} \qquad (3.2.15)$$

where f is the mean activity coefficient of the hydrogen and acetate ions. Here again the solutions are so dilute that the value of ρ is immaterial. The values of K_a in the last column of Table 3.1 are derived from the α' values and the limiting law for the activity coefficients; the value for the most concentrated solution is only increased by $\frac{1}{2}$ per cent if $\rho = 1\cdot 31$ $1^{\frac{1}{2}}$ mole$^{-\frac{1}{2}}$, which corresponds to $d = 4$ Å, is used in equation 3.2.15. There remains, however, a puzzling decrease of K_a over the concentration range 0·02 to 0·2M; no reasonable adjustment of ρ can explain it for if the values of α' are accepted the final value leads to $K_a = 1\cdot 750 \times 10^{-5}$ mole l^{-1} only if $\rho = 33$ mole$^{-\frac{1}{2}}$ l$^{\frac{1}{2}}$ which implies $d = 100$ Å. Empirically one finds that pK_a is an accurately linear function of $(1 - \alpha)c$, the concentration of undissociated acid. It has been suggested[13] that the effect may be due to the formation of acid dimers, which are well known in non-aqueous solvents. If the formation of dimers is wholly responsible for the effect, and a dimerization constant K_D is defined by $K_D = [(\text{HOAc})_2]/[\text{HOAc}]^2$, a value of K_a calculated from equation 3.2.14, which we now denote by K_a', is related to the true acidity constant K_a by

$$K_a' = \frac{K_a}{1 + 2\ K_D(1-\alpha)c} \qquad (3.2.16)$$

Provided the extent of dimerization is small, pK_a' should be a linear function of $(1 - \alpha)c$. The slope gives $K_D = 0\cdot 14$ l mole^{-1}, and this is reasonable.

Sometimes, particularly for non-aqueous solutions, there are insufficient data for completely dissociated salts to obtain $\varLambda_0$ independently. The calculation has then to be done for various values of $\varLambda_0$ and that value which gives the best constancy of K is assumed to be correct. Graphical procedures have been common; for instance combination of equations 3.2.4 and 3.2.14 gives

$$\frac{F}{\varLambda} = \frac{1}{\varLambda_0} + \frac{\varLambda c f^2}{\varLambda_0^2 F K_a} \qquad (3.2.17)$$

and iterative plotting of $F/\varLambda$ against $\varLambda cf^2/F$ leads to self-consistent values of $\varLambda_0$ and K_a. The advantages of calculating values of F from $F = 1 - Z$ (equation 3.2.10) have been emphasized[14] recently. Severe demands are made on the precision of the data and the method is unsuitable when K is less than about 10^{-5} mole l^{-1}.

The conductance method has been widely used for the precise determination of the acidity constants of uncharged weak acids, particularly carboxylic acids, in water, non-aqueous, and mixed solvents. It has been pre-eminent in the study of incomplete dissociation in solvents of low dielectric constant[15] of salts which are in the strong electrolyte category in water. Alkali metal and tetra-alkylammonium salts with univalent anions such as the halide ions, picrate, perchlorate, nitrate and thiocyanate have been a favourite topic of study in organic solvents such as ethylene chloride, pyridine, acetic acid, acetone, nitrobenzene, dimethyl formamide, acetonitrile and alcohols. Measurements have also been made in mixed solvents. Liquid ammonia and sulphur dioxide are two other solvents in which interesting studies of electrolytic dissociation have been made. In non-aqueous solvents it is usually necessary for both Λ_0 and K to be determined from the same data. If the dielectric constant is sufficiently low, as for example with pyridine which has a value of 12, even a salt such as potassium iodide has a dissociation constant of only 2×10^{-4} mole l^{-1} and the association is measurable in solutions sufficiently dilute for it to be immaterial whether the limiting law or an extension of it is used for the conductance of the free ions. This, however, is not true when the dielectric constant exceeds about twenty and for uni-univalent salts the dissociation constant then becomes sensitive to the value of d used in, say, equation 3.2.11. This difficulty is also encountered in studying the dissociation of many salts in water and will be discussed in the next section. In solvents with dielectric constants below about ten the existence of triple (at high concentrations quadruple) as well as double ions (ion pairs) becomes important.[16]

The conductance method has also been very useful in the study of colloidal electrolytes. For example, the sudden formation of micelles containing about fifty cations in a solution of cetylpyridinium bromide in water when the concentration exceeds about 0·001M causes large changes in conductivity and transport numbers.[17]

3.3. Strongly dissociated electrolytes

Some acids and many salts are strongly if not completely dissociated in water and other solvents of high dielectric constant. The equilibria are only amenable to study by the conductance method if measurable degrees of association are attained at concentrations at which Λ_i can be estimated with confidence. The higher the concentration the more Λ_i depends on the equation used and on the value of the d parameter. This will be illustrated with some examples taken from aqueous solution data.

For manganous oxalate[18] the value of α obtained by use of the limiting-law for Λ_i is 0·223 at 0·0075M, which is only 6 per cent higher than from the Leist equation with $d = 14$ Å. But for thallous chloride, for which small negative deviations[19, 20] from the limiting law (2½ per cent at 0·015M) are ascribed to incomplete dissociation, the position is very different. In Table 3.2 are compared values of $(1 - \alpha)$, the degree of association, calculated from the limiting law (α'), from the Leist equation (α'') and from the simplified Fuoss–Onsager equation (α'''). For the last two calculations d was taken as 3·57 Å. The differences

TABLE 3.2
Degree of association of thallous chloride in water at 25°C from conductance data
($\Lambda_0 = 151\cdot1$ int. Ω^{-1} cm^2 mole^{-1})

c mole l^{-1}	$(1 - \alpha')$ L.L.	$(1 - \alpha'')$ Leist	$(1 - \alpha''')$ F. and O.
0·00507	0·008	0·013	0·012
0·00604	0·010	0·016	0·014
0·00750	0·012	0·019	0·018
0·01000	0·019	0·027	0·026
0·01108	0·020	0·029	0·028
0·01501	0·025	0·038	0·037
0·01607	0·027	0·040	0·040

between α'' and α''' are insignificant, but the limiting-law values are about 50 per cent less throughout and will give a correspondingly different thermodynamic dissociation constant. In Fig. 3.1 are compared values of $(1 - \alpha)$ calculated for magnesium sulphate from the precise results of Dunsmore and James[21] using the Leist equation with $d = 4\cdot3$, 10 and 14 Å, and from the limiting law. The $(1 - \alpha)$ values are sensitive to d because $(1 - \alpha) = (\Lambda_i - \Lambda)/\Lambda_i$ and the differences $\Lambda_i - \Lambda$ are small and sensitive to d. Once the α values are accepted the d values used to calculate f, the ionic activity coefficient in $K = \alpha^2 c f^2 / (1 - \alpha)$ are less important. At the highest concentration the value of f^2 is 0·590 and 0·656 for $d = 0$ and $d = 14$ Å respectively. A logical procedure is to calculate f with a d value equal to that used to obtain α; in this way the three sets of K values in Table 3.3 are obtained. There is a trend in all three sets of values but that for K_2 and K_3 is small. The trend in K_2 and K_3 is eliminated if the Pitts conductance equation is used for Λ_i when $K_1 = 5\cdot14 \times 10^{-3}$ mole l^{-1}, $K_2 = 5\cdot45 \times 10^{-3}$ mole l^{-1}, $K_3 = 4\cdot91 \times 10^{-3}$ mole l^{-1}. The

Fuoss–Onsager equation gives much higher values of K_1 and K_2. This equation gives $K_1 = 8\cdot5 \times 10^{-3}$ mole l⁻¹, $K_2 = 5\cdot9 \times 10^{-3}$ mole l⁻¹, and $K_3 = 5\cdot0 \times 10^{-3}$ mole l⁻¹. The discrepancy suggests that for $d < 14$ Å the higher terms of the Fuoss–Onsager conductance equation take some account of the effects which can alternatively be described by introduction of a dissociation constant (i.e. the equation has some " built-in " provision for the effects of ion association). The reciprocal flexibility in the d and K parameters for magnesium sulphate is much

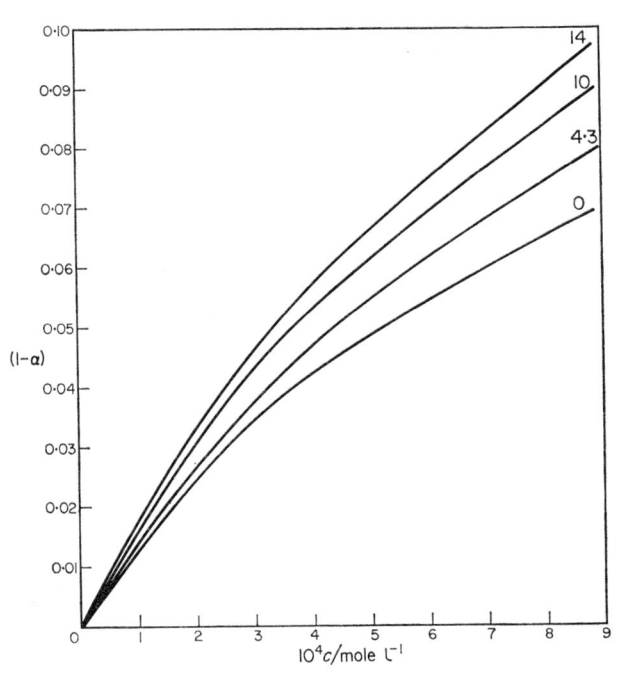

Fig. 3.1. Degree of association of magnesium sulphate in water at 25°C as a function of d in the Leist equation.

the same as with spectrophotometric measurements on copper sulphate (2.4). The reason for the flexibility is of course, similar; there is an appreciable population of magnesium and sulphate ions which can be treated to a good approximation as either free or paired, provided appropriate adjustments in the parameters are made. For a 3–3 electrolyte in water the relative population in the no-man's-land is much less. The value of α for $2\cdot40 \times 10^{-4}$M lanthanum cobalticyanide (25°C) is[22] 0·700 if the limiting law is used for Λ_i and only $2\tfrac{1}{2}$ per cent less if the Robinson–Stokes equation with $d = 32$ Å is substituted.

The values of large dissociation constants for salts are therefore often sensitive to an arbitrary choice of d. There are two further difficulties. Firstly, sometimes in water, and almost invariably in non-aqueous solvents, Λ_0 is not independently known. Secondly, if the electrolyte is unsymmetrical the limiting molar conductance of a complex species such as PbCl$^+$ has to be estimated. Whilst the reality of the incomplete dissociation of many salts in water cannot be denied,

TABLE 3.3
Dissociation constant of magnesium sulphate in water at 25°C from conductance data
$(\Lambda_0 = 133\cdot07$ int. Ω^{-1} cm^2 equiv.$^{-1})$
(K_1, K_2 and K_3 are for $\rho = 1\cdot415$ mole$^{-\frac{1}{2}}$ l$^{\frac{1}{2}}$ ($d = 4\cdot3$ Å), $\rho = 3\cdot290$ mole$^{-\frac{1}{2}}$ l$^{\frac{1}{2}}$ ($d = 10$ Å), and $\rho = 4\cdot606$ mole$^{-\frac{1}{2}}$ l$^{\frac{1}{2}}$ ($d = 14$ Å) respectively)

$10^4 c$ mole l^{-1}	$10^3 K_1$ mole l^{-1}	$10^3 K_2$ mole l^{-1}	$10^3 K_3$ mole l^{-1}
1·6366	5·16	4·79	4·64
1·994	5·23	4·74	4·48
3·090	5·37	4·80	4·51
4·270	5·50	4·96	4·66
5·597	5·60	5·00	4·69
6·006	5·71	5·05	4·79
7·197	5·74	5·18	4·80
8·380	5·79	5·17	4·84
8·846	5·78	5·20	4·88
	5·54 (mean)	4·99 (mean)	4·70 (mean)

and the general sequence of dissociation constants has been established, the values of the constants themselves are more arbitrary than for weaker electrolytes. Many of the values come from the work of C. W. Davies, Monk and their respective collaborators. Typical pK values of simple inorganic species in water, in addition to those already discussed, for which dissociation constants based on conductance measurements have been reported[23] are : KSO$_4^-$ (1·0), BaNO$_3^+$ (0·92), PbNO$_3^+$ (1·18), PbCl$^+$ (1·5), CdCl$^+$ (2·0), LaSO$_4^+$ (3·62), Co(NH$_3$)$_6^{3+}$SO$_4^{2-}$ (3·5), K$^+$Fe(CN)$_6^{3-}$ (1·3), K$^+$Fe(CN)$_6^{4-}$ (2·3), Ca^{2+}Fe(CN)$_6^{3-}$ (2·83), La^{3+}Fe(CN)$_6^{3-}$ (3·74). Salts of organic acids have also been investigated. Typical cases are the malonates of magnesium (2·8), nickel (4·1), copper (5·6), zinc (3·71) and cadmium (3·29). With these weaker electrolytes the pK values are less arbitrary. As a general rule pK values of less than about 3 are sensitive to the conductance equation used for the free ions.

Kay[11] has carefully re-analysed conductance data for alkali halides, nitrates, bromates and perchlorates in water, alcohols, alcohol + water mixtures and liquid ammonia. The simplified Fuoss–Onsager equation was used and Table 3.4 compares the dissociation constants for the alkali halides in ethanol obtained by Kay with those given by use of the limiting law. Another good example of the use of the simplified Fuoss–Onsager equation for aqueous solution data is provided by a study[24] of the association of KPF_6; the dissociation constant is 0·41 mole l⁻¹ at 25°C with $d = 4$ Å.

TABLE 3.4
Dissociation constants of alkali halides in ethanol at 25°C

Salt	$10^2 K$ mole l⁻¹ L.L.	$10^2 K$ mole l⁻¹ F. and O.	d Å F. and O.
LiCl	1·80	3·7	4·4
NaCl	1·25	2·27	4·0
KCl	0·80	1·05	4·6

3.4. Mixed electrolytes

Conductimetric measurements are ideally suited to the precise study of the dissociation of neutral species in the absence of other electrolytes. They are not in general suited to the study of equilibria in the presence of other non-reacting ions. Sometimes, however, measurements on mixtures can be useful. Large deviations from additivity at constant total ionic strength can provide qualitative or semi-quantitative evidence for incomplete dissociation.[25] The high mobility of the hydrogen ion in water has been exploited[26] to study the hydrolytic equilibria of ions such as Fe^{3+} and Bi^{3+} at a high acidity in a 3M perchlorate medium. In such a medium the conductivity increases approximately linearly as hydrogen ions are substituted for sodium ions up to 0·6M, the increase being 0·1 per cent for a change of 0·001M in [H⁺]. The conductivity can therefore be used as a sensitive measure of [H⁺] in a solution containing a trivalent ion in hydrolytic equilibrium.

REFERENCES

1. ARRHENIUS, S., *Z. phys. Chem.*, 1887, **1**, 631.
2. MACINNES, D. A. and SHEDLOVSKY, T., *J. Amer. chem. Soc.*, 1932, **54**, 1429.
3. ONSAGER, L., *Physik. Z.*, 1927, **28**, 277.

4. FUOSS, R. M., *J. Amer. chem. Soc.*, 1935, **57**, 488.
5. SHEDLOVSKY, T., *J. Franklin Inst.*, 1938, **225**, 739.
6. LEIST, M., *Z. phys. Chem. (Leipzig)*, 1955, **205**, 16.
7. PITTS, E., *Proc. Roy. Soc.*, 1953, **A.217**, 43.
8. FUOSS, R. M. and ONSAGER, L., *J. phys. Chem.*, 1957, **61**, 668 ; FUOSS, R. M. and ONSAGER, L., *J. phys. Chem.*, 1963, **67**, 621.
9. ROBINSON, R. A. and STOKES, R. H., *J. Amer. chem. Soc.*, 1954, **76**, 1991.
10. STOKES, R. H., *J. phys. Chem.*, 1961, **65**, 1242.
11. KAY, R. L., *J. Amer. chem. Soc.*, 1960, **82**, 2099.
12. PRUE, J. E. and SHERRINGTON, P. J., *Trans. Faraday Soc.*, 1961, **57**, 1795.
13. KATCHALSKY, A., EISENBERG, H. and LIFSON, S., *J. Amer. chem. Soc.*, 1951, **73**, 5889.
14. WIRTH, H., *J. phys. Chem.*, 1961, **65**, 1441.
15. FISCHER, L., WINKLER, G. and JANDER, G., *Z. Elektrochem.*, 1958, **62**, 1.
16. FUOSS, R. M. and KRAUS, C. A., *J. Amer. chem. Soc.*, 1933, **55**, 2387 ; KRAUS, C. A., *J. phys. Chem.*, 1956, **60**, 129.
17. HARTLEY, G. S., COLLIE, B. and SAMIS, C. S., *Trans. Faraday Soc.*, 1936, **32**, 795.
18. MONEY, R. W. and DAVIES, C. W., *Trans. Faraday Soc.*, 1932, **28**, 609.
19. GARRETT, A. B. and VELLENGA, S. J., *J. Amer. chem. Soc.*, 1945, **67**, 225.
20. BRAY, W. C. and WINNINGHOF, W. J., *J. Amer. chem. Soc.*, 1931, **33**, 1663.
21. DUNSMORE, H. S. and JAMES, J. C., *J. chem. Soc.*, 1951, 2925.
22. GUGGENHEIM, E. A. and PRUE, J. E., *Physicochemical Calculations*, North-Holland, Amsterdam, 1956, p. 366.
23. DAVIES, C. W., *The Structure of Electrolytic Solutions*, Ed. HAMER, W. J., John Wiley, New York, 1959, p. 19.
24. ROBINSON, R. A., STOKES, J. M. and STOKES, R. H., *J. phys. Chem.*, 1961, **65**, 542.
25. DAVIES, C. W., *Endeavour*, 1945, **4**, 114.
26. BERECKI-BIEDERMANN, C. and BIEDERMANN, G., " Int. Conf. Co-ordination Chem.", *Chem. Soc. Special Publ.*, 1959, No. 13, p. 190.

CHAPTER 4

ELECTROCHEMICAL CELLS*

4.1. Introduction

The simplest kind of electrochemical cell is one in which the solutions surrounding each of the two electrodes and all the intervening solution are so similar in composition that they can be regarded as identical, except with respect to the reactions at the electrodes. As an example consider the cell

$$H_2, Pt \mid HCl(aq) \mid HgCl \mid Hg$$

with the cell reaction

$$\tfrac{1}{2}H_2(g) + HgCl(s) \rightarrow H^+(aq) + Cl^-(aq) + Hg(l).$$

The solubilities of hydrogen and of mercurous chloride in water are too low significantly to affect the compositions of the solutions. The e.m.f. is given by

$$E = E^\ominus - \frac{RT}{F} \ln [H^+][Cl^-]\gamma_H\gamma_{Cl} \qquad (4.1.1)$$

where $E^\ominus$ is the standard potential for the cell and [ ] and γ are here used to denote molalities and activity coefficients (on the molality scale); the activities of pure solid or liquid substances are constant and all measurements are made at or corrected to a constant pressure of hydrogen of one atmosphere. Measurements on this particular cell over a range of concentration of hydrochloric acid can be used to calculate both the standard potential $E^\ominus$ and values of the activity coefficient of hydrochloric acid (assumed completely dissociated) over the relevant concentration range. Substituting in equation 4.1.1 for $\gamma_H\gamma_{Cl}$ from the equation corresponding to 1.3.4 on the molality scale

$$-\log \gamma_i = z_i^2 A' I'^{\tfrac{1}{2}}/(1 + \rho' I'^{\tfrac{1}{2}}) - B_i' I' \qquad (4.1.2)$$

and setting $\rho' = 1$, one obtains

$$E + \frac{2RT}{F} \left\{ \ln [HCl] - \frac{A'I'^{\tfrac{1}{2}}}{1 + I'^{\tfrac{1}{2}}} \right\} = E^\ominus - B'I' \qquad (4.1.3)$$

---

* The sign convention which will be used for the electromotive force of cells is that recommended[1] by the International Union of Pure and Applied Chemistry in 1953.

With the best available data,[2] a plot of the left-hand side of equation 4.1.3 against the concentration of hydrochloric acid gives a straight line from which individual points seldom deviate by as much as 0·01 mV with $E^\ominus = 267\cdot96$ mV, $B' = 0\cdot234$ kg mole^{-1}.

4.2. Equilibrium constants of electron-transfer reactions

In so far as it is possible satisfactorily to ensure that the e.m.f. of a cell is entirely due to the electrode reactions and to take account of activity coefficients, the calomel electrode in the cell discussed in 4.1 may be replaced by any other reversible electrode and the corresponding $E^\ominus$ determined. The standard e.m.f. of the half-cell $Zn^{2+} \mid Zn$ is the e.m.f. of the cell

$$H_2, Pt \mid H^+ \mid\mid Zn^{2+} \mid Zn$$

implying the cell reaction

$$\tfrac{1}{2}H_2 + \tfrac{1}{2}Zn^{2+} \rightarrow H^+ + \tfrac{1}{2}Zn$$

and with each solute species present at an activity equal to 1 mole kg^{-1} and the hydrogen at a partial pressure of one atmosphere.

These $E^\ominus$ values for aqueous solution at 25°C are called standard electrode potentials, e.g. $E^\ominus(Zn^{2+} \mid Zn) = -0\cdot76$ V, and are available in tables.[3] The algebraic difference between any two such values gives the electromotive force of a new cell and from this the equilibrium constant of the corresponding electron-transfer (redox) reaction is calculated by $FE^\ominus = RT \ln K$. For example, from $E^\ominus(Ag^+ \mid Ag) = +0\cdot80$ V and $E^\ominus(Zn^{2+} \mid Zn) = -0\cdot76$ V the $E^\ominus$ value of the cell

$$Ag \mid Ag^+ \mid\mid Zn^{2+} \mid Zn$$

with the cell reaction

$$Ag + \tfrac{1}{2}Zn^{2+} \rightarrow Ag^+ + \tfrac{1}{2}Zn$$

is $-0\cdot76 - 0\cdot80$ V $= -1\cdot56$ V and

$$K = \frac{[Ag^+]\gamma_{Ag}}{[Zn^{2+}]^{\frac{1}{2}}\gamma_{Zn}^{\frac{1}{2}}} = 10^{-26\cdot4} \text{ mole}^{\frac{1}{2}} \text{ kg}^{-\frac{1}{2}}$$

Without formally writing down the cell and its reaction it is clear from the sign and magnitude of the original standard electrode potentials that zinc has a greater tendency to go into solution by the process $\tfrac{1}{2}Zn \rightarrow \tfrac{1}{2}Zn^{2+} + e^-$ than silver by the process $Ag \rightarrow Ag^+ + e^-$ and that therefore the equilibrium constant for the reaction as written must be less than one.

The more powerful a reducing agent a system is, the more negative is the standard electrode potential. Rather than what are properly called electrode potentials, the quantity sometimes tabulated has the opposite sign, being the e.m.f. of the cell with the hydrogen electrode on the right, again by convention regarded as positive when the hydrogen electrode is positive.

Study of protolytic equilibria by measurement of $[H^+]\gamma_H\gamma_{Cl}$

The cell

$$H_2, Pt \mid HCl(aq) \mid HgCl \mid Hg$$

or the corresponding cell with the calomel electrode replaced by a

TABLE 4.1
Acidity constant of $H_2PO_4^-$ in water at 25°C

$10^3 a$ mole kg^{-1}	$10^3 b$ mole kg^{-1}	$10^3 c$ mole kg^{-1}	E mV	$p([H^+]\gamma_H\gamma_{Cl})$	pK^{app}
5·983	2·915	9·721	786·72	6·754	7·206
8·245	4·017	13·395	777·05	6·732	7·202
9·812	4·780	15·942	772·10	6·724	7·206
11·493	5·599	18·672	767·40	6·713	7·207
13·579	6·615	22·061	762·24	6·697	7·204
18·140	8·837	29·472	753·45	6·675	7·206
19·238	9·372	31·256	751·65	6·670	7·206
20·878	10·171	33·920	749·22	6·665	7·208
22·211	10·820	36·085	747·26	6·658	7·207
26·496	12·908	43·048	741·78	6·642	7·207

silver–silver chloride electrode is capable of giving very precise results. Provided that a solution does not contain anything that interferes with the electrode reactions or affects the concentration of free chloride ions, and that $E^\ominus$ is known, the quantity $[H^+]\gamma_H\gamma_{Cl}$ can be calculated by inserting measured values of E in equation 4.1.1. Individual electrodes are conveniently standardized[4] (and errors arising from the difficulty of exactly reproducing the physical state of solids thereby avoided) by making a measurement on a 0·01 molal solution of hydrochloric acid for which $\gamma_\pm = 0·904$ at 25°C. For instance, interpolation from published data[5] gives $E = 509·83$ mV as the e.m.f. of the cell discussed in 4.1 for 0·01 molal hydrochloric acid at 25°C, from which $E^\ominus = 268·00$ mV. Table 4.1 gives some values of the e.m.f. at 25°C of the cell

$$H_2, Pt \mid KH_2PO_4(a), KNaHPO_4(b), NaCl(c) \mid HgCl \mid Hg$$

measured by the same worker.[6] These together with $E^{\ominus} = 268{\cdot}00$ mV lead to the values of $[H^+]\gamma_H\gamma_{Cl}$ shown in the fifth column of the table, and thence to $R[H^+]\gamma_H\gamma_{Cl} = K\gamma_{Cl}\gamma_{H_2PO_4}/\gamma_{HPO_4}$ where $R = b/a$ is the buffer ratio and K the acidity constant of $H_2PO_4^-$. If the equation

$$\log \frac{\gamma_{Cl}\gamma_{H_2PO_4}}{\gamma_{HPO_4}} = \frac{2A'I'^{\frac{1}{2}}}{1 + I'^{\frac{1}{2}}} - B'I' \qquad (4.3.1)$$

is substituted for the activity coefficients then

$$pK^{app} = pR + p([H^+]\gamma_H\gamma_{Cl}) + 2A'I'^{\frac{1}{2}}/(1 + I'^{\frac{1}{2}}) = pK + B'I' \qquad (4.3.2)$$

and a plot of the left-hand side against I should give a straight line with an intercept on the ordinate axis of pK. In fact, inspection of the final column of the table shows that $B' = 0$ within the experimental precision, and $pK = 7{\cdot}206$. (The slight discrepancy with the result of $7{\cdot}200$ recorded in the original paper is due to the use there of $\rho' = 1{\cdot}247$ kg$^{\frac{1}{2}}$ mole$^{-\frac{1}{2}}$ rather than 1 kg$^{\frac{1}{2}}$ mole$^{-\frac{1}{2}}$ in equation 4.3.1 and to slight differences in the values assumed for $E^{\ominus}$ and the fundamental constants.)

For acetic acid the same method gives[7] $K_m = 1{\cdot}75_1 \times 10^{-5}$ mole kg^{-1} at 25°C. This is in satisfactory agreement with the value $K_m = K_c/d_o = 1{\cdot}74_9 \times 10^{-5}$ mole l^{-1}/$0{\cdot}9971$ kg l^{-1} = $1{\cdot}75_4$ mole kg^{-1} obtained by the conductimetric method (3.1). For acetate buffers there is sufficient reliable information about specific interaction coefficients (equation 1.3.6) to calculate independently the values of $\gamma_{Cl}\gamma_{HOAc}/\gamma_{OAc}$ needed to convert $R[H^+]\gamma_H\gamma_{Cl}$ values to K, and it is satisfying that the value of K obtained in this way[8] agrees within experimental error with that obtained by the simple method illustrated above for $H_2PO_4^-$. One hopes that such a check will eventually be possible in many other cases. Measurements on the type of cell discussed in this section have attained a pre-eminent position for the determination of precise values of acidity constants of both organic acids and bases and inorganic species such as oxyacids and aquocations. The ionization constant of water itself can be obtained by measuring $[H^+]\gamma_H\gamma_{Cl}$ in solutions of alkali metal hydroxide of known $[OH^-]$ values. The method can also be used for measurements in non-aqueous and mixed solvents. Determinations can conveniently be made over a temperature range. The hydrogen electrode can be replaced by a glass electrode if the effects of the changing asymmetry potential is eliminated.[9]

In the example discussed in detail, the buffer ratio $[HPO_4^{2-}]/[H_2PO_4^-]$ was equated to the ratio of the stoichiometric concentrations. This is no longer valid if the pK of the acid is far removed from 7 and the buffer ratio is then $(b + [H^+])/(a - [H^+])$ for acids of low pK and $(b - [OH^-])/(a + [OH^-])$ for those of high pK. In the determination of the pK of an acid such as HSO_4^-, therefore, the buffer ratio at any concentration can only be calculated with sufficient accuracy after $[H^+]$ has been calculated from $[H^+]\gamma_H\gamma_{Cl}$, which requires that ρ' be arbitrarily fixed in an equation such as

$$-\log \gamma_H\gamma_{Cl} = 2A'I'^{\frac{1}{2}}/(1 + \rho'I'^{\frac{1}{2}}) \qquad (4.3.3)$$

TABLE 4.2

Acidity constant of HSO_4^- in water at 25°C
(R_1, R_2 and R_3 are for $\rho' = 1 \cdot 314$ mole$^{-\frac{1}{2}}$ kg$^{\frac{1}{2}}$ (d = 4 Å), $\rho' = 1 \cdot 643$ mole$^{-\frac{1}{2}}$ kg$^{\frac{1}{2}}$ (d = 5 Å), and $\rho' = 2 \cdot 629$ mole$^{-\frac{1}{2}}$ kg$^{\frac{1}{2}}$ (d = 8 Å) respectively.)

$10^3 m_1$ mole kg^{-1}	$10^3 m_2$ mole kg^{-1}	$E - E^\ominus$ mV	$10^3[H^+]\gamma_H\gamma_{Cl}$ mole kg^{-1}	R_1	R_2	R_3
1·2711	1·3498	319·04	3·184	3·912	3·688	3·209
2·1606	2·2481	293·71	5·020	2·644	2·452	2·070
3·5560	2·9684	272·80	6·884	1·996	1·842	1·477
4·6073	5·4429	255·93	10·245	1·460	1·288	0·825
6·7748	5·4093	242·76	11·630	1·294	1·122	0·825
8·1637	8·4059	230·92	15·302	1·026	0·867	0·613

In Table 4.2 the effect of the choice of ρ' on R is shown for some of the results of Nair and Nancollas[10] for the cell

$$H_2, Pt \mid HCl(m_1), H_2SO_4(m_2) \mid AgCl \mid Ag$$

The same values of ρ' have been used in equation 4.3.2 to calculate for all Nair and Nancollas' results the values of pK^{app} plotted in Fig. 4.1. The effect of ρ' at this stage of the calculation is much less important than its effect on R; even for the highest ionic strength pK^{app} is only changed by 0·023 by a change of ρ' from 1·643 mole$^{-\frac{1}{2}}$ kg$^{\frac{1}{2}}$ to $\rho' = 2 \cdot 629$ mole$^{-\frac{1}{2}}$ kg$^{\frac{1}{2}}$. The plots of pK^{app} against I' for the different values of R do not extrapolate to a common intercept and there is no reason why they should. It is only if one demands that the results be fitted by a *single* adjustable parameter in addition to K, i.e. that the same value of ρ' be used in equations 4.3.2 and 4.3.3 and that B' in equation 4.3.2

be zero, that pK can be unambiguously determined. This is equivalent to Nair and Nancollas' choice of the equation

$$-\log \gamma_i = A'z_i^2 \{I'^{\frac{1}{2}}/(1 + I'^{\frac{1}{2}}) - 0\cdot 20 I'\} \qquad (4.3.4)$$

for all ionic activity coefficients. This corresponds approximately to equation 4.3.3 with $\rho' = 1\cdot 41$ mole$^{-\frac{1}{2}}$ kg$^{\frac{1}{2}}$, and the pK value of $1\cdot 96$ reported by Nair and Nancollas is therefore different from that of $1\cdot 99$ obtained from $\rho' = 1\cdot 64$ mole$^{-\frac{1}{2}}$ kg$^{\frac{1}{2}}$ (R_2 in Fig. 4.1).

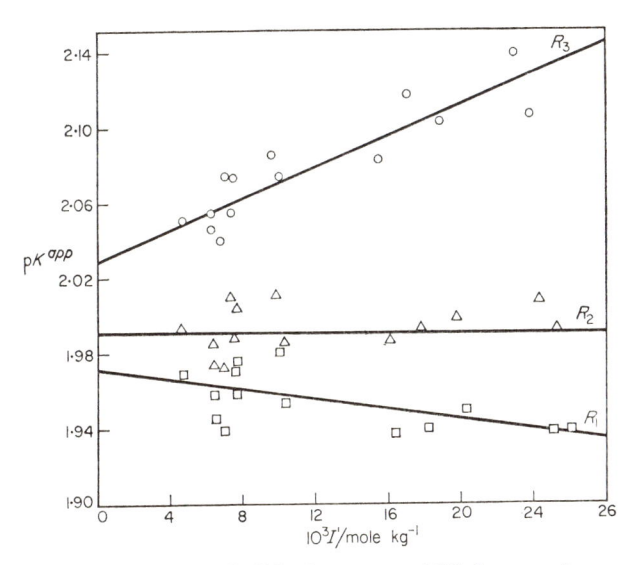

Fig. 4.1. The acidity constant of HSO_4^- in water at 25°C from e.m.f. measurements.

A parallel problem is encountered[11] in the determination of the dissociation constant of the base $CaOH^+$. Eventually such uncertainties may be reduced by an increased knowledge of specific interaction coefficients (equation 1.3.6) and by enhanced experimental precision reducing the range of parameters which fit the results (note the similar problems discussed in 2.4 and 3.3). A way of circumventing the problem in some cases[12] is to combine spectrophotometrically determined values of the buffer ratio with the values of $[H^+]\gamma_H\gamma_{Cl}$.

4.4. Thermodynamic stability constants of metal ion complexes from cells

The number of cells from which precise e.m.f. data of the kind so far discussed can be obtained is limited by the scarcity of well-behaved electrodes. Suitable electrodes, however, exist for the study of electrolytes such as the halides of zinc, cadmium, lead and thallium, and the sulphates of zinc, cadmium, copper and thallium. For instance, several precise sets of measurements have been made on the cell

Pb (two-phase Hg amalgam) | $PbCl_2$(aq) | AgCl | Ag

with and without added alkali metal chloride.[13] For this cell

$$E = E^\ominus - (RT/2F)\ln [Pb^{2+}][Cl^-]^2 \gamma_{Pb}\gamma_{Cl}^2 \qquad (4.4.1)$$

The mean ionic activity coefficients (stoichiometric activity coefficients) determined from this equation with the aid of a provisional value of $E^\ominus$ and by assuming lead chloride to be completely dissociated are markedly less than those of an alkaline earth chloride and do not show Debye–Hückel behaviour in their concentration dependence. This suggests that lead chloride is incompletely dissociated; if only one complex $PbCl^+$ exists then

$$E = E^\ominus - (RT/2F)\ln (m - [PbCl^+])(2m - [PbCl^+])^2 \gamma_{Pb}\gamma_{Cl}^2 \qquad (4.4.2)$$

where m is the concentration of lead chloride and the activity coefficients now refer to free ions. The dissociation constant of $PbCl^+$ is given by

$$\begin{aligned} K &= (m - [PbCl^+])(2m - [PbCl^+])\gamma_{Pb}\gamma_{Cl}/[PbCl^+]\gamma_{PbCl} \\ &= Q\gamma_{Pb}\gamma_{Cl}/\gamma_{PbCl} \end{aligned} \qquad (4.4.3)$$

Unfortunately in order to determine values of Q which can be extrapolated to give K it is necessary to know both $E^\ominus$ and $\gamma_{Pb}\gamma_{Cl}^2$ for the free ions. It is in turn very difficult to make a reliable extrapolation to get $E^\ominus$ unless K is known. There are three adjustable parameters, viz. $E^\ominus$, K and d which have to be obtained from the same set of experimental data. The numerical significance of the best-fitting set of parameters is uncertain even when obtained from the best available experimental results. The parameters obtained from e.m.f. measurements on bi-bivalent sulphates are similarly uncertain[14]; because of the lack of independent knowledge of $E^\ominus$, even the stoichiometric activity coefficients are unreliable.

The uncertainty about $E^\ominus$ can be eliminated at the cost of making measurements in solutions of mixed electrolytes if the ligand has basic properties. Complex formation by the basic species of a buffer with an added cation has an effect on the buffer ratio R which can be determined by e.m.f. measurement on cells of the type discussed in 4.3.

For example, for the HSO_4^-/SO_4^{2-} buffers it was shown that up to an ionic strength of 0·026 mole kg^{-1} results for the cell

$$H_2, Pt \mid HCl, H_2SO_4 \mid AgCl \mid Ag$$

could be described by

$$pK = pR + p([H^+]\gamma_H\gamma_{Cl}) + 2A'I'^{\frac{1}{2}}/(1 + \rho'I'^{\frac{1}{2}}) \quad (4.4.4)$$

with $pK = 1·99$ and $\rho' = 1·314$ mole$^{-\frac{1}{2}}$ kg$^{\frac{1}{2}}$. For the cell

$$H_2, Pt \mid HCl(m_1), MgSO_4(m_2) \mid AgCl \mid Ag$$

TABLE 4.3
Dissociation constant of magnesium sulphate in water at 25°C from e.m.f. measurements
$\rho' = 1·643$ mole$^{-\frac{1}{2}}$ kg$^{\frac{1}{2}}$ $(d = 5$ Å$)$

$10^3 m_1$ mole kg^{-1}	$10^3 m_2$ mole kg^{-1}	$E - E^\ominus$ mV	$10^3[H^+]\gamma_H\gamma_{Cl}$ mole kg^{-1}	$10^3 K$ mole kg^{-1}
5·522	3·619	277·04	3·759	5·5
7·127	5·222	265·66	4·535	5·8
7·575	5·148	262·49	4·827	6·1
6·170	5·333	273·13	3·917	6·6
4·392	38·497	307·09	1·467	5·7
3·652	16·727	308·38	1·678	5·5

it is then reasonable to use experimental values of $p([H^+]\gamma_H\gamma_{Cl})$ to determine the buffer ratio $R = [SO_4^{2-}]/[HSO_4^-]$ from equation 4.4.4. With the help of the relations

$$[HSO_4^-] = m_1 - [H^+] \quad (4.4.5)$$

$$[Mg^{2+}] = [HSO_4^-] + [SO_4^{2-}] \quad (4.4.6)$$

$$[MgSO_4] = m_2 - [Mg^{2+}] \quad (4.4.7)$$

it is then possible to calculate values of the dissociation constant of magnesium sulphate from the equation

$$pK(MgSO_4) = p([Mg^{2+}][SO_4^{2-}]/[MgSO_4]) + 8A'I'^{\frac{1}{2}}/(1 + \rho'I'^{\frac{1}{2}}) \quad (4.4.8)$$

Values calculated from the results of Nair and Nancollas[15] are shown in Table 4.3. The average value of K is $5·90 \times 10^{-3}$ mole kg^{-1} which is in good agreement with a value of $5·5 \times 10^{-3}$ mole l^{-1} interpolated for the same value of ρ' from the conductance results (3.3). The value obtained by Nair and Nancollas who used equation 4.3.4 for activity coefficients is $5·6 \times 10^{-3}$ mole kg^{-1}.

This method is of general applicability and has been used to determine the dissociation constants of species such as transition metal oxalates and malonates,[16] and the magnesium salts of phosphoric acids of biological importance.[17] Such complexes are more stable than those with the sulphate ion and the acids themselves are weaker; the pK values are much less sensitive to arbitrary assumptions about activity coefficients. However, uncertainties about ionic activity coefficients and values of $E^{\ominus}$ have often forced those interested in the determination of successive stability constants for step-wise complex formation to abandon the attempt to determine thermodynamic stability constants and to content themselves with the determination of equilibrium quotient values in some swamping and supposedly inert background electrolyte such as 3M sodium perchlorate.

4.5. Use of swamping ionic media

There are two distinct purposes in the use of a concentrated salt solution rather than water as the solvent medium. The first is to maintain activity coefficients constant and so permit the use of the equilibrium equation in its classical form. Unfortunately, there is little knowledge of the limits within which this is true. At high ionic concentrations constancy of ionic strength is, of course, no assurance that activity coefficients are constant, because specific effects may become dominant. For example, the activity coefficient of hydrochloric acid is 1·316 in its own solution at 3M, but 1·070 at zero concentration in 3M sodium chloride.[18] It would be of great interest to have some measurements of the variation of the activity coefficients of salts present at low concentrations in supposedly swamping media. The change in activity coefficient with concentration of a salt AB due to specific interactions between A and B will probably[19] be roughly the same whether the solvent is pure water or a 3M solution of a salt XY. If this is so, known specific interaction coefficients for simple uni-univalent salts[20] suggest that an increase of concentration of AB from 0 to 0·1M in a swamping medium might easily cause a change in γ_{AB} of 5 per cent, and the effect could be much larger with highly charged ions. Of course, no information can be obtained about chemical equilibria involving an ion of the swamping electrolyte any more than it can about equilibria involving the solvent.

The cell

$$\text{H}_2, \text{Pt} \mid \text{H}^+, \text{Cl}^- \ldots \mid \text{AgCl} \mid \text{Ag}$$

has been widely used under swamping conditions, often with glass or hydroquinone electrodes substituted for the hydrogen electrode. The e.m.f. is assumed to be given by $E = E^\circ - (RT/F) \ln [\text{H}^+][\text{Cl}^-]$ where the standard potential E° now refers to unit activity of hydrogen and chloride ions in the swamping medium and not in water; the system can be calibrated to give a direct measure of hydrogen ion concentration. The ease and precision with which measurements can be made with this type of cell make it suitable not only for the study of protolytic equilibria but also for the study of complex ion equilibria. Provided the equilibrium quotient of the acid HL is known for the medium in question and its actual concentration is known (this is frequently assumed to be known with sufficient accuracy by making measurements with a high stoichiometric concentration), the free ligand concentration in the solution, whatever other equilibria involving L$^-$ are also involved, is given by $[\text{L}^-] = Q[\text{HL}]/[\text{H}^+]$.

The second purpose for which a swamping medium is often used is to make the solution around an electrode sufficiently similar in total ionic composition with that of the solution with which it is in contact for this junction to be neglected as a source of changes in the e.m.f. of the cell. For a cell such as

M | M(ClO$_4$)$_n$, NaClO$_4$(3M) | NaClO$_4$(3M) | Reference electrode

the e.m.f. can be written

$$E = (RT/xF) \ln [\text{M}^{x+}] + E_{\text{ref}} \qquad (4.5.1)$$

provided the changes in composition are not too large. The cell is directly calibrated to give values of [M^{x+}]. The effect of additions of a ligand which forms complexes with M^{x+} is then examined; such experiments are conveniently done as a titration and a series of experimental points thereby obtained.

The extent to which swamping media adequately fulfil their twofold function is much less certain than one would wish; a large and ever-increasing number of investigations of complex ion and hydrolytic equilibria of cations are made with such media. It is only in the cases of simple equilibria that it is sometimes[21] possible to obtain unambiguous values of equilibrium quotients. Consider the cell

H$_2$, Pt | NH$_3$(c_1), NH$_4$Cl(c_2) | KCl(3·5M) | HCl(c), KCl($C - c$) | Pt, H$_2$
                KCl($C - c_2$)

Values of [H$^+$] are obtained from the ideal equation $E = (RT/F)\ln[\text{H}^+]/c$

and provisional values of the equilibrium quotient $Q = [NH_3][H^+]/[NH_4^+]$ are then calculated over a range of values of c_2 and c which are kept in a constant ratio to one another. A plot of log Q against c gives a linear plot which can be extrapolated to $c = 0$. When the two electrode solutions are identical errors due to differences in the two solutions must disappear. Values of the equilibrium quotient can, of course, be measured over a range of ionic strengths and the thermodynamic acidity constant obtained therefrom.

4.6. The calculation of successive stability constants

The removal of successive protons from a polybasic acid usually proceeds in clearly defined steps and the titration curve has a characteristic wave-like form; for instance the successive pK values of polybasic oxyacids are separated by units of about 5. By contrast the successive pK's of complexes of monodentate ligands with cations usually differ by less than one unit. This means that there is a pronounced overlap of the equilibria and solutions inevitably contain several species in appreciable concentrations. For example, at a concentration of 10^{-2}M free ammonia there is[22] present in a solution of nickel nitrate 4 per cent Ni^{2+}, 25 per cent $Ni(NH_3)^{2+}$, 44 per cent $Ni(NH_3)_2^{2+}$, 23 per cent $Ni(NH_3)_3^{2+}$ and 4 per cent $Ni(NH_3)_4^{2+}$. This complicates the calculation of the corresponding stability constants from experimental data, for it is seldom possible to select concentration regions in which single equilibria are of over-riding importance. These complications have resulted in the appearance of an excessively voluminous literature dealing with the various methods, graphical and otherwise, of solving the relevant sets of simultaneous equations.

A quantity which has been found generally useful in the analysis of data is the ligand number $\bar{n}$ defined by

$$\bar{n} = (C_L - [L])/C_M \tag{4.6.1}$$

where C_L and C_M are the stoichiometric concentration of ligand and metal cation. Let us consider a set of successive equilibria between the species M, ML, ML$_2$, ML$_3$, ..., ML$_n$. We define the overall stability quotients for the formation of individual complexes by

$$\beta_n = [ML_n]/[M][L]^n \tag{4.6.2}$$

Then

$$\bar{n} = \frac{C_L - [L]}{C_M} = \frac{\sum_{n=1}^{N} n[ML_n]}{[M] + \sum_{n=1}^{N} [ML_n]} = \frac{\sum_{n=1}^{N} n\beta_n[L]^n}{1 + \sum_{n=1}^{N} \beta_n[L]^n} \tag{4.6.3}$$

Rearrangement of this equation gives

$$\frac{\bar{n}}{(1-\bar{n})[L]} = \beta_1 + \beta_2 \frac{(2-\bar{n})}{(1-\bar{n})}[L] + \sum_{n=3}^{N} \beta_n \frac{(n-\bar{n})}{(1-\bar{n})}[L]^{n-1} \quad (4.6.4)$$

If the free ligand concentration is measurable, $\bar{n}$ is directly calculable, and if $N = 2$ a linear plot gives β_1 as intercept and β_2 as slope. Such a plot is shown in Fig. 4.2 using values of [L] and $\bar{n}$ calculated[23] from some measurements of J. Bjerrum on the system $Ag^+ + 2NH_3$. If $N > 2$ one method of establishing approximate values of the constants

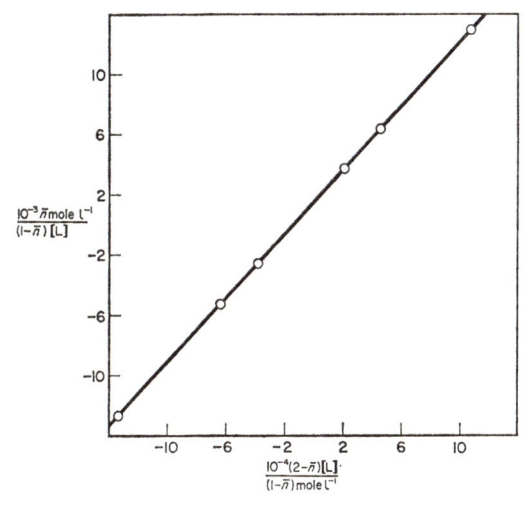

FIG. 4.2. Graphical method for determining stability quotients of $Ag(NH_3)^+$ and $Ag(NH_3)_2^+$.

is to use values of $\bar{n} < 2$ to obtain β_1 and β_2, and then to use these constants to calculate a function which when plotted against $[L]^2$ for $2 < \bar{n} < 4$ gives β_3 as intercept and β_4 as limiting slope, and so on. A good deal of effort has been devoted to developing curve-fitting and other methods for evaluating the best values of the constants in such cases;[24] in future the use of electronic computers for such purposes will no doubt become standard practice.

It is to be hoped that the availability of computers will result in a more thorough examination than hitherto of the agreement of calculated and observed experimental quantities, and the extent to which both a postulated set of equilibria are confirmed and the individual values of a set of parameters are uniquely determined by the data.

If it is the free metal ion concentration that is measured we have

$$C_M/[M] = 1 + \sum_{n=1}^{N} \beta_n[L]^n \qquad (4.6.6)$$

At sufficiently low ligand concentrations a plot of $(C_M - [M])/[M][L]$ against $[L]$ will give an intercept of β_1 and a limiting slope of β_2; this procedure as in the case of equation 4.6.4 can be extended. It is only in special cases that $[L]$ can be equated to or easily calculated from the stoichiometric concentration.

The analysis of results is particularly complicated when polynuclear complexes of the general formula M_mL_n arise, which is often so when the ligand is OH^-, e.g. the main hydrolysis product of Bi^{3+} is $Bi_6(OH)_{12}^{6+}$. In such cases, e.g. Bi^{3+}, UO_2^{2+}, Zr^{4+}, even qualitative conclusions about the major species present have needed revision in the light of more direct evidence from ultra-centrifuge work and of more precise e.m.f. measurements.[25] In e.m.f. studies determination of the concentrations of both the ligand and the metal ion is highly desirable.

If the ligand is polydentate the method of analysis of the results is in principle the same, but the detail is often simplified because the stability constants are larger and more widely separated. Further simplification can be achieved in some cases by making measurements with an at least tenfold excess of cation over ligand. For instance, if the polyamine cation $N(-CH_2 \cdot CH_2 \cdot NH_3^+)_3$ is titrated[26] against hydroxide in the presence of a tenfold excess of Co^{2+}, Cu^{2+}, Zn^{2+}, Cd^{2+} the reaction is

$$N(-CH_2 \cdot CH_2 \cdot NH_3^+)_3 + M^{2+} \rightleftarrows N(-CH_2 \cdot CH_2 \cdot NH_2)_3 M^{2+} + 3H^+$$

which is formally the same, if $[M^{2+}]$ is effectively constant, as that of a tribasic acid which loses all its protons in one step. In other cases (Ni^{2+}, Hg^{2+}), however, intermediate protonated complexes occur and in one case (Ag^+) polynuclear complexes are formed.

4.7. Polarographic measurement of cation concentration

The potential of a dropping mercury electrode can also be used as a measure of free metal ion concentration. If the cathodic reaction is

$$M^{x+} + Hg + xe^- \rightarrow M(Hg)$$

the e.m.f. for a reversible reaction is

$$E = E^\ominus - \frac{RT}{xF} \ln \frac{[M]_{Hg}}{[M^{x+}]_s} \qquad (4.7.1)$$

where for a given medium $E^{\ominus}$ depends on the nature of M, the reference electrode and any junction potentials. Activity coefficients are omitted because experiments are necessarily made with a supporting electrolyte in about one-hundred fold excess which carries practically all the current. $[M]_{Hg}$ denotes the concentration of M in the amalgam at the surface of a drop of mercury and $[M^{x+}]_s$ is the concentration of M^{x+} in the solution around the drop. The change in e.m.f. $\Delta E = E' - E$ if a ligand is added which forms complexes with the cation (and does not itself produce an overlapping polarographic wave) will be

$$\Delta E = E' - E = \frac{RT}{xF} \ln \frac{[M]_{Hg}}{[M]'_{Hg}} + \frac{RT}{xF} \ln \frac{[M^{x+}]'_s}{[M^{x+}]_s} \qquad (4.7.2)$$

It is reasonable to assume that the concentration of metal in the amalgam at the surface of a mercury drop is proportional to the current, so $[M]_{Hg}/[M]'_{Hg} = i/i'$. With a supporting electrolyte that carries all the current, metal ions only reach the electrode by diffusion. The current in the absence of ligands will by the simplest assumption be given by

$$i = k_M([M^{x+}] - [M^{x+}]_s) \qquad (4.7.3)$$

where k_M depends on the diffusion coefficient of M^{x+} and the capillary characteristics. The greatest possible current (the diffusion current) will be attained when the surface concentration of M^{x+} is negligible compared with that in the bulk of the solution so

$$i_D = k_M[M^{x+}] \qquad (4.7.4)$$

and

$$\frac{[M^{x+}]_s}{[M^{x+}]} = \frac{i_D - i}{i_D} \qquad (4.7.5)$$

Substitution in equation 4.7.2 gives

$$\Delta E = \frac{RT}{xF} \ln \frac{i}{i'} + \frac{RT}{xF} \ln \frac{[M^{x+}]'_s}{[M^{x+}]} + \frac{RT}{xF} \ln \frac{i_D}{i_D - i} \qquad (4.7.6)$$

It is therefore possible to measure the equilibrium concentration $[M^{x+}]'_s$ of free metal ions in the solution around a mercury drop. To calculate the stability quotients of complexes, it is usual to assume that the ligand concentration in this region is equal to that in the bulk solution. Many assumptions are necessary in the polarographic method but it is of general applicability and has been much used to obtain approximate values of stability constants for complexes of zinc, cadmium, lead, copper(II), tin(II) and thallium(I) with both organic and inorganic ligands.[24]

REFERENCES

1. I.U.P.A.C. *Manual of Physicochemical Symbols and Terminology*, Butterworths, London, 1954, p. 4.
2. HILLS, G. J. and IVES, D. J. G., *J. chem. Soc.*, 1951, 315.
3. LATIMER, W. M., *The Oxidation States of the Elements and their Potentials in Aqueous Solutions*, Prentice-Hall, 1952, pp. 340–345.
4. BATES, R. G., et al., *J. chem. Phys.*, 1956, **25**, 361.
5. GRZYBOWSKI, A. K., *J. phys. Chem.*, 1958, **62**, 550.
6. GRZYBOWSKI, A. K., *J. phys. Chem.*, 1958, **62**, 555.
7. GUGGENHEIM, E. A. and PRUE, J. E., *Physicochemical Calculations*, North-Holland, Amsterdam, 1956, p. 360.
8. PITZER, K. S. and BREWER, L., *Thermodynamics* by LEWIS, G. N. and RANDALL, M., 2nd ed., McGraw-Hill, New York, 1961, p. 587; KING, E. J., *Acid–base Equilibria*, Pergamon Press, Oxford 1965, Chap. 3.
9. KING, E. J. and PRUE, J. E., *J. chem. Soc.*, 1961, 275; ZIELEN, A. J., *J. phys. Chem.*, 1963, **67**, 1474.
10. NAIR, V. S. K. and NANCOLLAS, G. H., *J. chem. Soc.*, 1958, 4144.
11. BATES, R. G., BOWER, V. E., CANHAM, R. G. and PRUE, J. E., *Trans. Faraday Soc.*, 1959, **55**, 2062.
12. BATES, R. G. and SCHWARZENBACH, G., *Helv. chim. acta*, 1954, **37**, 1069.
13. GARRELS, R. M. and GUCKER, F. T., *Chem. Rev.*, 1949, **44**, 117.
14. GUGGENHEIM, E. A., *Disc. Faraday Soc.*, 1957, **24**, 53.
15. NAIR, V. S. K. and NANCOLLAS, G. H., *J. chem. Soc.*, 1958, 3706.
16. MCAULEY, A. and NANCOLLAS, G. H., *J. chem. Soc.*, 1961, 4367.
17. CLARKE, H. B., CUSWORTH, D. C. and DATTA, S.P., *Biochem. J.*, 1954, **58**, 146.
18. HARNED, H. S., *J. phys. Chem.*, 1959, **63**, 1299.
19. PITZER, K. S. and BREWER, L., *Thermodynamics* by LEWIS, G. N. and RANDALL, M., 2nd ed., McGraw-Hill, New York, 1961, p. 577.
20. GUGGENHEIM, E. A. and TURGEON, J. C., *Trans. Faraday Soc.*, 1955, **51**, 747.
21. EVERETT, D. H. and WYNNE-JONES, W. F. K., *Proc. Roy. Soc.*, 1938, **A.169**, 190.
22. BJERRUM, J., *Metal Ammine Formation in Aqueous Solution*, Haase, Copenhagen, 1941, p. 189.
23. GUGGENHEIM, E. A. and PRUE, J. E., *Physicochemical Calculations*, North-Holland, Amsterdam, 1956, p. 360.
24. ROSSOTTI, F. J. C. and ROSSOTTI, H., *The Determination of Stability Constants*, McGraw-Hill, New York, 1961.
25. RUSH, R. M., JOHNSON, J. S. and KRAUS, K. A., *Inorg. Chem.*, 1962, **1**, 378; BAES, C. F. and MEYER, N. J., *Inorg. Chem.*, 1962, **1**, 780.
26. PRUE, J. E. and SCHWARZENBACH, G., *Helv. chim. acta*, 1950, **33**, 963.

CHAPTER 5

SOLUBILITY AND DISTRIBUTION

5.1. Solubility

Experimental values of solubilities and distribution ratios are influenced by ionic equilibria involving the species that is partitioned. From such measurements equilibrium constants can be determined.

In the solubility method, the solubility product relation is used to calculate, from the stoichiometric concentration of an ion A in a solution saturated with a salt AB, the concentration of the second ion B in the same solution. As an example of the application of this to the determination of a dissociation constant consider the analysis of the results in Table 5.1. The slight solubility c_{Ag}(obs) of silver chloride in

TABLE 5.1
Dissociation constants of $AgCl$ *and* $AgCl_2^-$ *in water at 25°C*

10^3[NaCl] mole l^{-1}	$10^6 c_{Ag}$(obs) mole l^{-1}	10^6[Ag$^+$] mole l^{-1}	$10^6 c_{Ag}$(calc) mole l^{-1}	$10^6 \Delta c_{Ag}$ mole l^{-1}
0 0592	3·31	3·04	3·39	+0·08
0·112	2·04	1·62	1·97	−0·07
0·551	0·692	0·339	0·69	0·00
1·10	0·525	0·173	0·54	+0·02
2·75	0·490	0·072	0·49	0·00
5·50	0·575	0·038	0·54	−0·03
11·0	0·660	0 020	0·67	+0·01
27·0	1·10	0·0030	1·13	+0·03
55·0	1·95	0 0050	1·92	−0·03

aqueous solutions of sodium chloride has been accurately measured radiochemically[1] by using silver chloride labelled with the ^{110}Ag isotope by neutron irradiation. Conductimetric and electrometric methods give[2] a consistent value of $1·769 \times 10^{-10}$ mole^2 l^{-2} for the solubility product $P = $ [Ag$^+$][Cl$^-$]f^2. Setting

$$-\log f = AI^{\frac{1}{2}}/(1 + I^{\frac{1}{2}}) \qquad (5.1.1)$$

and assuming that the chloride ion concentration is equal to the stoichiometric concentration of sodium chloride, the values of the free silver ion concentration given in column three of Table 5.1 are obtained. The discrepancy between the values of c_{Ag}(obs) and [Ag$^+$] must be due to the presence in the solution of complex species. For the first five points in Table 5.1 the difference between the total dissolved silver concentration and that of silver ions fluctuates about an average value of $0·34 \times 10^{-6}$ mole l^{-1} which suggests the presence of this amount of dissolved AgCl, but at higher concentrations the difference increases which suggests that AgCl$_2^-$ is then formed. Values of the total dissolved silver concentration can be calculated from

$$c_{Ag} = [Ag^+] + [AgCl] + [AgCl_2^-] \tag{5.1.2}$$

$$c_{Ag} = \frac{P}{[Cl^-]f^2} + \frac{P}{K_{AgCl}} + \frac{P}{K_{AgCl_2^-}}[Cl^-] \tag{5.1.3}$$

where K_{AgCl} and $K_{AgCl_2^-}$ are dissociation constants defined by $K_{AgCl} = [Ag^+][Cl^-]f^2/[AgCl]$ and $K_{AgCl_2^-} = [Ag^+][Cl^-]^2f^2/[AgCl_2^-]$. From $K_{AgCl} = 5·2 \times 10^{-4}$ mole l^{-1}, $K_{AgCl_2^-} = 6·2 \times 10^{-6}$ mole^2 l^{-2} and equation 5.1.1 for f the values of c_{Ag}(calc) in Table 5.1 are obtained. The differences $\Delta c_{Ag} = c_{Ag}$(calc) $- c_{Ag}$(obs) are satisfactorily small. The best-fitting values of the dissociation constants are insensitive to the denominator used in the activity coefficient equation; the first term on the right-hand side of equation 5.1.3 has fallen to 3 per cent of the total at an ionic strength of 0·01 mole l^{-1}. That silver chloride is a weak electrolyte is a surprise to many; in many contexts the AgCl can be ignored because its concentration remains constant in saturated solution. The radiochemical method has also been used to study the formation of the even stabler complexes of silver ions with bromide and iodide ions, and of polynuclear complexes of silver and halide;[3] the latter was done by investigating the solubility of silver halides in silver nitrate solution with isotopically labelled halogen.

In the next example, the dissociation constant in contrast to constants for silver halide complexes, is sensitive to the assumption made about the activity coefficient of free ions. Bell and George[4] determined by volumetric analysis the iodate concentration c_{IO_3} in potassium chloride solutions saturated with thallous iodate. The results are given in Table 5.2. If it is valid to assume that the free iodate ion concentration is equal to its stoichiometric concentration the thallous ion concentration in the solution can be calculated from

$[\text{Tl}^+][\text{IO}_3^-]f^2 = P$. For the solubility product P a value $3 \cdot 069 \times 10^{-6}$ mole^2 l^{-2} is used, obtained by extrapolation[5] of Bell and George's results for potassium chloride and potassium thiocyanate, and for f the equation

$$-\log f = AI^{\frac{1}{2}}/(1 + \rho I^{\frac{1}{2}}) \tag{5.1.4}$$

with two alternative values of ρ of 1 mole$^{-\frac{1}{2}}$ l$^{\frac{1}{2}}$ and $1 \cdot 5$ mole$^{-\frac{1}{2}}$ l$^{\frac{1}{2}}$. The values of f and $[\text{Tl}^+]$ calculated therefrom by successive approximations are designated by subscripts 1 and 2 respectively in Table 5.2.

TABLE 5.2
Dissociation constant of thallous chloride in water at 25°C

$10^3[\text{KCl}]$ mole l^{-1}	$10^3 c_{\text{IO}_3}$ mole l^{-1}	$10^3[\text{Tl}^+]_1$ mole l^{-1}	$10^3[\text{Tl}^+]_2$ mole l^{-1}	f_1^2	f_2^2
4·90	1·930	1·903	1·890	0·836	0·841
12·57	2·025	1·953	1·929	0·776	0·786
25·65	2·158	1·988	1·948	0·715	0·730
40·81	2·266	2·031	1·966	0·667	0·689
54·22	2·359	2·046	1·966	0·636	0·662

$10^3[\text{TlCl}]_1$ mole l^{-1}	$10^3[\text{TlCl}]_2$ mole l^{-1}	$10 Q_1$ mole l^{-1}	$10 Q_2$ mole l^{-1}	$10 K_1$ mole l^{-1}	$10 K_2$ mole l^{-1}
0·027	0·040	3·43	2·31	2·87	1·94
0·072	0·096	3·38	2·49	2·62	1·96
0·170	0·201	2·99	2·36	2·14	1·74
0·235	0·300	3·50	2·66	2·34	1·83
0·313	0·393	3·52	2·70	2·24	1·79

Even though the maximum difference between $[\text{Tl}^+]_1$ and $[\text{Tl}^+]_2$ is 4 per cent, the value of $[\text{TlCl}]$ calculated from

$$[\text{TlCl}] = c_{\text{Tl}} - [\text{Tl}^+] = c_{\text{IO}_3} - [\text{Tl}^+] \tag{5.1.5}$$

is such a small difference between large quantities that the corresponding maximum difference between $[\text{TlCl}]_1$ and $[\text{TlCl}]_2$ exceeds 20 per cent. The conductimetric method for determining $[\text{TlCl}]$ in pure thallous chloride solution had similar uncertainties (3.3). The uncertainty about the activity coefficient of the free ions is of much less importance in converting values of the equilibrium quotient $Q = [\text{Tl}^+][\text{Cl}^-]/[\text{TlCl}]$ to the thermodynamic equilibrium constant $K = Qf^2$. The average values of K in Table 5.2 of $0 \cdot 244$ and $0 \cdot 185$ mole l^{-1}

bracket the value reported by Bell and George of 0·215 mole l⁻¹. Bell and George used

$$-\log f = AI^{\frac{1}{2}}/(1 + I^{\frac{1}{2}}) - 0·20\,I \qquad (5.1.6)$$

for the activity coefficients and also applied small corrections for the incomplete dissociation of thallous and potassium iodates.

Smith[6] has suggested an ingenious way for finding the supposedly constant concentration of undissociated thallous chloride in its saturated aqueous solutions. This depends on comparative solubility measurements in solutions of two chlorides of different valency type. Let the solubilities of thallous chloride be s_A and s_B in concentrations c_A and c_B of chlorides ACl and BCl₂, both assumed to be completely dissociated. Then

$$I_A = c_A + s_A - [\text{TlCl}] \qquad (5.1.7)$$

$$I_B = 3c_B + s_B - [\text{TlCl}] \qquad (5.1.8)$$

A pair of " corresponding " solutions is located for which $c_A + s_A = 3c_B + s_B$, so that $I_A = I_B$. Provided that the mean ionic activity coefficient of thallous chloride is a function only of the ionic strength in the two solutions,

$$[\text{Tl}^+]_A[\text{Cl}^-]_A = [\text{Tl}^+]_B[\text{Cl}^-]_B \qquad (5.1.9)$$

$$(I - c_A)I = (I - 3c_B)(2c_B + I - 3c_B) \qquad (5.1.10)$$

Equation 5.1.10 is solved for I and [TlCl] then calculated from equation 5.1.7. Published data on the solubility of thallous chloride in solutions of 1–1 and 2–1 chlorides show an absence of specific cation effects which supports the validity of equation 5.1.9, but more data are needed before Smith's suggestion can be thoroughly tested. The method could also be applied to comparative measurements of solubility in solutions of MCl and MX. Such measurements with NaCl and NaClO₄ have recently been reported by Macaskill and Panckhurst.[7] In their method of analysing the results they find the value of [TlCl] that is consistent with f in $[\text{Tl}^+][\text{Cl}^-]f^2 = P$ having the same dependence on ionic strength in the two solutions. They report a dissociation constant of 0·30 mole l⁻¹ with f given by equation 5.1.4 with $\rho = 1$ mole$^{-\frac{1}{2}}$ l$^{\frac{1}{2}}$. They also analyse their results using independent information about the specific ionic interaction coefficients between thallous and perchlorate and between chloride and sodium ions. They then obtain $K = 0·23$ mole l⁻¹ with f derived from equation 1.3.6 with

$\rho = 1$ mole$^{-\frac{1}{2}}$ l$^{\frac{1}{2}}$ and $B_{\text{Tl, Cl}} = 0\cdot07$ mole^{-1} l. The dependence of the dissociation constant of thallous chloride on the assumptions made about the activity coefficients of free ions is evident.

The solubility method has been widely used to determine the dissociation constants of complex ions; sparingly soluble iodates such as those of calcium, barium, copper(I), silver, cerium(III), and thallium(I) are particularly convenient.

Protolytic equilibria have also been studied, e.g. pK for iodic acid and for the bisulphate ion have been calculated from the effect of pH on the solubilities of silver iodate and silver sulphate. The method is also useful for sparingly soluble acids. The electrolyte solutions are of necessity mixed and uncertainties about activity coefficients and subsidiary equilibria can be serious. Measurements are often made in a constant swamping medium in an effort to circumvent the activity coefficient problem, particularly if there is step-wise complex formation. Scandinavian workers have been particularly active in this field. For example, the complexes of silver with thiosulphate ions have been studied[8] in 4M sodium perchlorate by using $NaAgS_2O_3 \cdot H_2O$ as the sparingly soluble salt.

5.2. Distribution method

This method uses a knowledge of the partition coefficient of a species between two phases in order to calculate from its concentration in one phase its unknown concentration in a second phase. It was one of the first methods to be used for the study of complex formation in aqueous solutions. Generally speaking, only neutral species are extracted from aqueous solutions into organic solvents, and complexes with such species as NH_3, I_2 and $HgCl_2$ were amongst those first studied. The accuracy of the method is limited by such problems as the mutual solubility of the two solvents and uncertainties about activity coefficients in the organic phase. Ingenious variants of the straightforward procedure have sometimes been used to overcome these difficulties. For example,[9] analysis of a solution of iodine in pure water after equilibration through the vapour phase with a solution of iodine containing iodide ions gives the concentration of iodine in the solution containing I_3^-. More recently the same technique has been used[10] to obtain free bromine concentrations and thence, with the help of the stability constants of polyhalide ions, bromide ion concentrations in solutions containing mercury(II). Accurate values for the stability constants of the cupric ammines were obtained[11] by studying by a transpiration method the effects of cupric ions on the partial pressure of ammonia above its aqueous solution.

The use by atomic energy workers of organic solvents for the extraction of nuclear fission products from aqueous solutions has given renewed impetus to distribution studies. As a method of studying equilibria in aqueous solution the usefulness of the conventional method has been extended by the greater availability of radioactive isotopes. Accurate measurements can be made on very dilute solutions and the method has real advantages in the study of protolytic and complex ion equilibria if species are slightly soluble or if polymerization occurs at higher concentrations. The uncharged species which distributes itself between the two phases may be one of the complexes under study or it may be formed from an auxiliary ligand. Consider the case of a complex ML'_2 of an auxiliary ligand L' with the cation M, which distributes itself between an organic phase and water with a partition coefficient $P = [ML'_2]_0/[ML'_2]$. The distribution ratio R of M between organic and aqueous phases when the latter phase also contains a ligand L which forms a complex ML insoluble in the organic phase is given by

$$R = \frac{[ML'_2]_0}{[M] + [ML'] + [ML'_2] + [ML]} \qquad (5.2.1)$$

The assumption that only a single complex ML and no mixed complexes are formed by L is obviously not of general validity. Manning and Monk[12] studied the distribution of tracer amounts (ca. 10^{-6} mole l^{-1}) of ^{60}Co as $CoCl_2$ between aqueous sodium oxalate, adjusted to a pH of 6 (with a trace of sodium acetate) and made up to an ionic strength I with sodium chloride, and an organic phase of 0·001M oxine (8-hydroxyquinoline) in chloroform. The results are given in Table 5.3. The complex extracted is $Co(oxine)_2$. If equation 5.2.1 is valid for this case

$$R = \frac{P\beta'_2[L']^2}{1 + \beta'_1[L'] + \beta'_2[L']^2 + \beta_1[L]} = \frac{P\beta'_2[L']^2}{1 + \Delta + \beta_1[L]} \qquad (5.2.2)$$

where β's are overall stability quotients, and Δ is a constant if the same concentration of auxiliary ligand is present throughout. If R^0 is the value of R in the absence of L

$$\frac{R^0}{R} - 1 = \frac{\beta_1[L]}{1+\Delta} \qquad (5.2.3)$$

If Δ is assumed to be negligible the values of β_1 calculated at four ionic strengths are shown in Table 5.3. [L] is equated to the stoichiometric concentration. If then the mean ionic activity coefficient of cobalt oxalate is calculated from

$$-\log f = 4AI^{\frac{1}{2}}/(1 + \rho I^{\frac{1}{2}}) \qquad (5.2.4)$$

with $\rho = 1\cdot 742$ mole$^{-\frac{1}{2}}$ l$^{\frac{1}{2}}$, the values of the thermodynamic dissociation constant shown in the table are obtained from $K = \beta_1^{-1} f^2$.

TABLE 5.3

Dissociation constant of cobaltous oxalate in water at 25°C by the distribution method

I mole l^{-1}	10^5[L] mole l^{-1}	R	$10^{-4}\beta_1$ l mole^{-1}	$10^5 K$ mole l^{-1}
0·02	0	49·15	—	
	8·00	22·49	1·48	
	10·00	19·71	1·49	2·30
	12·00	17·52	1·50	
0·04	0	117·5	—	
	8·00	68·31	1·05	
	9·00	59·72	1·07	2·32
	10·00	56·84	1·07	
0·08	0	92·31	—	
	7·00	61·90	0·702	
	7·50	59·68	0·729	2·34
	8·00	58·28	0·730	
0·10	0	11·14	—	
	8·00	7·39	0·634	
	10·00	6·76	0·648	2·28
	12·00	6·27	0·647	

The average value of $2\cdot 3 \times 10^{-5}$ mole l^{-1} is in only fair accord with a value of $1\cdot 6 \times 10^{-5}$ mole l^{-1} obtained[13] by e.m.f. measurements on the cell

glass electrode | NaHC$_2$O$_4$, Na$_2$C$_2$O$_4$, CoCl$_2$, NaCl | AgCl | Ag

However, the agreement is improved if some account is taken of Δ. 10 ml samples of the two phases were equilibrated, and the partition coefficient of oxine between chloroform and water is[13] about 370 so an approximate value of [HL′] is $0\cdot 5 \times 10^{-3}$ mole l^{-1} $\times$ 370^{-1} = $1\cdot 25 \times 10^{-6}$ mole l^{-1}. The acidity constant of oxine is[14] $1\cdot 55 \times 10^{-10}$ mole l^{-1} and β_1' is approximately $1\cdot 25 \times 10^9$ l mole^{-1} so at a pH of 6

$$\Delta \simeq \beta_1'[L] \simeq \beta_1' K_{HL'}[HL'][H^+]^{-1} \qquad (5.2.5)$$
$$= 1\cdot 25 \times 10^9 \times 1\cdot 55 \times 10^{-10} \times 1\cdot 25 \times 10^{-6} \times 10^6 = 0\cdot 24$$

The corrected value of K is $2\cdot 3 \times 10^{-5}$ mole$^{-1}/(1+\varDelta) = 1\cdot 8 \times 10^{-5}$ mole l^{-1}. This ignores any variation of $\varDelta$ with ionic strength. It may also be that the correction should be further adjusted to take account of slight CoOAc$^+$ complex formation.

It will be apparent that equilibrium constants obtained by the distribution method are not usually as accurate as those obtained by the methods described in earlier chapters. However, with radiochemical analysis values of useful accuracy are rather easily and quickly obtained. The method has been much used to determine stability quotients for complexes of actinide cations. Popular auxiliary ligands have been alkylphosphates, octylamine, nitrosonaphthols, and other organic chelating agents.

5.3. Ion-exchange method

In the example in the preceding section the effect of an added ligand on the distribution ratio of a cation between an organic phase and the aqueous phase was investigated. Instead it is possible to study the distribution of the cation between a cation-exchange resin and the aqueous phase. If it is a sodium resin, and the cation is M^{2+}, the exchange reaction is

$$2Na^+_R + M^{2+} \rightleftarrows 2Na^+ + M^{2+}_R$$

A partition coefficient and a distribution ratio are defined by

$$P = \frac{[M^{2+}]_R}{[M^{2+}]} = K_E \Pi(f) \frac{[Na^+]_R^2}{[Na^+]^2} \tag{5.3.1}$$

$$R = \frac{[M]_R}{[M]} \tag{5.3.2}$$

where K_E is an exchange equilibrium constant and $\Pi(f)$ an activity coefficient factor. If a ligand L present in the aqueous phase forms only a single complex ML, and neither L nor ML enters the resin,

$$R = \frac{[M]_R}{[M^{2+}]+[ML]} = \frac{P[M^{2+}]}{[M^{2+}]+\beta_1[M^{2+}][L]} = \frac{P}{1+\beta_1[L]} \tag{5.3.3}$$

or

$$\frac{R^0}{R} - 1 = \beta_1[L] \quad \text{(cf. equation 5.2.3)} \tag{5.3.4}$$

where $R^0 = P$ is the value of R when $[L] = 0$.

The radiotracer technique makes it possible to study the effect of an added ligand on R under conditions where changes in [Na⁺] and [Na⁺]$_R$ and in the activity coefficient factor $\Pi(f)$ are so small that P remains effectively constant. The distribution ratio R of ⁶⁰Co in tracer amounts (ca. 10^{-6} mole l⁻¹) as $CoCl_2$ between sodium chloride solutions of ionic strength 0·16 mole l⁻¹ and a sodium-loaded sulphonated polystyrene resin was affected as shown[15] in Table 5.4 by the addition

TABLE 5.4

Dissociation constant of cobaltous oxalate in aqueous sodium chloride solution ($I = 0·16$ mole l⁻¹) at 25°C by the ion-exchange method

10^3[L] mole l⁻¹	R	$10^{-3}\beta_1$ l mole⁻¹
0	5·75	—
0·25	2·39	5·6
0·50	1·46	5·8
1·00	0·780	6·3
2·00	0·361	7·4

of sodium oxalate (NaL). The pH of the solutions was adjusted to 5·5. The values of β_1 calculated from equation 5.3.4 are not very constant but if the average value of $6·3 \times 10^{-3}$ mole l⁻¹ is converted to a dissociation constant using equation 5.2.4 for f, $K = \beta_1^{-1} f^2 = 1·7 \times 10^5$ mole l⁻¹. The agreement with the other methods (5.2) is probably fortuitously good.

The ion-exchange method has been extended to cases where more than one cationic species from solution enters the resin, and some work has been done with anion exchangers. Spectrophotometry and polarography have been used for analysis as well as radiochemical methods. The method is not a precise one, but like the distribution method is capable of giving results of useful accuracy.

REFERENCES

1. JONTE, J. H. and MARTIN, D. S., *J. Amer. chem. Soc.*, 1952, **74**, 2052.
2. GUGGENHEIM, E. A. and PRUE, J. E., *Trans. Faraday Soc.*, 1954, **50**, 231.
3. LIESER, K. H., *Z. anorg. Chem.*, 1960, **304**, 296.
4. BELL, R. P. and GEORGE, J. H. B., *Trans. Faraday Soc.*, 1953, **49**, 619.
5. GUGGENHEIM, E. A. and PRUE, J. E., *Physicochemical Calculations*, North-Holland, Amsterdam, 1956, p. 228.
6. SMITH, G. F., *Trans. Faraday Soc.*, 1962, **58**, 350.
7. MACASKILL, J. B. and PANCKHURST, M. H., *Austral. J. Chem.*, 1964, **17**, 522.

8. Nilsson, R. O., *Arkiv. Kemi*, 1958, **12**, 219.
9. Jones, G. and Kaplan, B. B., *J. Amer. chem. Soc.*, 1928, **50**, 1600.
10. Scaife, D. B. and Tyrrell, H. J. V., *J. chem. Soc.*, 1958, 392.
11. Bjerrum, J., *Kgl. Danske Videnskab. Selskab, Mat.-fys. Medd.*, 1931, **11**, No. 5.
12. Manning, P. G. and Monk, C. B., *Trans. Faraday Soc.*, 1961, **57**, 1996.
13. McAuley, A. and Nancollas, G. H., *J. chem. Soc.*, 1961, 4367.
14. King, E. J., *Acid–base Equilibria*, Pergamon Press, 1965, p. 128.
15. Schubert, J., Lind, E. L., Westfall, W. M., Pfleger, R. and Li, N. C., *J. Amer. chem. Soc.*, 1958, **80**, 4799.

CHAPTER 6

COLLIGATIVE PROPERTIES

6.1. Introduction

In an ideal solution, a measurement of the vapour pressure or a related property of the solvent provides a measure of the total concentration of solute particles from which the dissociation constant of a dissociated solute can be calculated. For example, if the depression of the freezing-point of water is θ and the cryoscopic constant λ, the total concentration of solute particles $\Sigma[S_i]$ is then given for very dilute solutions by $\Sigma[S_i] = \theta\lambda^{-1}$. In fact, as has been appreciated for over half a century, in ionic solutions the behaviour of solute particles is far from ideal and due allowance has to be made for the effects of long-range coulombic forces on θ even in very dilute solutions. Furthermore, severe demands are made with respect to experimental accuracy. For measurements on dilute solutions the highest accuracy is obtained by the cryoscopic method.

6.2. Cryoscopy

The cryoscopic constant for water is small ($\lambda = 1\cdot 86$ deg mole^{-1} l) and even with an accuracy of about 10^{-4} deg in θ useful measurements cannot be made below a solute concentration of about $0\cdot 01$ mole l^{-1}. The method is therefore unsuited to the study of the dissociation of weak electrolytes (the degree of dissociation of a complex is roughly 10 per cent at $c = 100\ K$), whilst if an electrolyte is only appreciably associated at ionic strengths above $0\cdot 1$ mole l^{-1} allowance for the effect of interionic forces on θ is difficult. There is therefore only a narrow class of intermediate electrolytes the dissociation of which is conveniently investigated by the cryoscopic method. As an example, consider some measurements on bi-bivalent sulphates[1] and in particular the results for magnesium sulphate given in Table 6.1. Quantitatively

$$\theta(1 + b\theta)\lambda^{-1} = m(1 - \alpha) + 2\alpha\phi \qquad (6.2.1)$$

where α is the degree of dissociation, $\lambda = 1\cdot 860$ deg mole^{-1} kg and $b = 4\cdot 8 \times 10^{-4}$ deg^{-1}. It will be seen that $b\theta$ can be neglected for the values of θ in Table 6.1. Undissociated molecules are assumed to behave ideally, and ϕ is an osmotic coefficient which takes account of the non-ideal behaviour of free ions (it is equal to ϕ' of the original paper). It is related by the Gibbs–Duhem equation to γ, the mean activity coefficient of the free ions. We have

$$\phi = 1 - \tfrac{1}{3}\alpha'_D z^2 I'^{\frac{1}{2}} \sigma(\rho' I'^{\frac{1}{2}}) \qquad (6.2.2)$$

and values of the function $\sigma(y)$ are available in tables.[2] I' is given by $I' = 4\alpha m$ and after a value of ρ' has been chosen successive approximations are necessary to obtain α from equations 6.2.1 and 6.2.2.

TABLE 6.1
Dissociation constant of magnesium sulphate in water at 0°C from cryoscopic measurements

(α_1, α_2 and α_3 are from $\rho' = 2 \cdot 793$ mole$^{-\frac{1}{2}}$ kg$^{\frac{1}{2}}$ ($d = 8 \cdot 6$ Å), $\rho' = 1 \cdot 949$ mole$^{-\frac{1}{2}}$ kg$^{\frac{1}{2}}$ ($d = 6$ Å) and $\rho' = 1 \cdot 396$ mole$^{-\frac{1}{2}}$ kg$^{\frac{1}{2}}$ ($d = 4 \cdot 3$ Å) respectively.)

m mole kg^{-1}	θ deg	α_1	α_2	α_3	$10^3 K_1$ mole kg^{-1}	$10^3 K_2$ mole kg^{-1}	$10^3 K_3$ mole kg^{-1}
0·00610	0·0176	0·734	0·776	0·816	5·1	6·1	7·6
0·01523	0·0406	0·620	0·693	0·771	4·9	6·3	8·7
0·01935	0·0505	0·589	0·662	0·767	4·9	6·1	9·6
0·02650	0·0672	0·540	0·625	0·741	4·6	5·8	9·1
0·02830	0·0715	0·536	0·622	0·743	4·7	5·9	9·5
0·03210	0·0803	0·521	0·621	0·744	4·7	5·9	10·0
				Mean	4·8	6·0	9·1

Values of α for three values of ρ' are given in Table 6.1, together with values of the dissociation constant calculated from these α values by the equations

$$K = \alpha^2 m \gamma^2 / (1 - \alpha) \qquad (6.2.3)$$

$$-\log \gamma = A' z^2 I'^{\frac{1}{2}} / (1 + \rho' I'^{\frac{1}{2}}) \qquad (6.2.4)$$

Clearly the same value of ρ' must be used in equation 6.2.4 as in 6.2.2. The values of K are reasonably constant, but different, whichever of the three values of ρ' is used. Sufficient faith in the experimental accuracy suggests a preference for K_2 on the grounds of larger systematic trends in K_1 and K_3, but just as with conductimetric measurements it will be seen that there is an appreciable population of magnesium and sulphate ions which can be treated to a good approximation either as free or paired provided that ρ' and K are adjusted accordingly. Similar results are obtained for several other bi-bivalent sulphates. The values of K are sensitive to the values of ρ' assumed in calculating them, and increase by about 50 per cent as ρ' changes from 4·6 mole$^{-\frac{1}{2}}$ kg$^{\frac{1}{2}}$ ($d = 13·9$ Å) to 1·6 mole$^{-\frac{1}{2}}$ kg$^{\frac{1}{2}}$ ($d = 5$ Å). The values for a fixed ρ' are, however, very similar and Table 6.2 gives some values for $\rho' = 2·4$ mole$^{-\frac{1}{2}}$ kg$^{\frac{1}{2}}$ ($d = 6·9$ Å). This one expects from the similarity of the thermodynamic behaviour of these sulphates; the molecular reason for this is that the complexes are probably almost entirely outer-sphere (8.4).

TABLE 6.2

Dissociation constants for bi-bivalent sulphates in water at 0°C
($\rho' = 2·4$ mole$^{-\frac{1}{2}}$ kg$^{\frac{1}{2}}$)

	CaSO$_4$	MgSO$_4$	CoSO$_4$	NiSO$_4$	CuSO$_4$	ZnSO$_4$
$10^3 K$/mole kg^{-1}	4·3	5·2	5·5	5·2	4·1	5·5

It is interesting to note here that the results can alternatively be fitted without introducing the concept of an equilibrium between free and associated ions by choosing a suitable value (3·5–4·0 Å) for the ionic diameter a (the distinction between the closest distances of approach of free and associated ions now disappears), and carrying through a numerical integration of the Poisson–Boltzmann equation.[3] This takes account of the higher terms in the equation which are neglected in the derivation of equation 6.2.2. The treatment assumes among other things that ionic interactions right up to the closest distance of approach a are between non-polarizable spheres in a medium of dielectric constant equal to that of the solvent. The incomplete dissociation treatment does not make this assumption, but it will be interesting to examine later whether the K values are consistent with such a picture. Treatments which include the higher terms of the Poisson–Boltzmann equation, and also that of Mayer,[4] may be loosely said to have " built-in " provision for ion association. A corresponding treatment of conductivity is possible, but complicated.

It has been shown earlier that spectrophotometric (2.4) conductimetric (3.3) and e.m.f. (4.4) measurements on bi-bivalent sulphates are conveniently treated in terms of incomplete dissociation, and it is important to compare the results of the earlier methods with those from cryoscopy. Results for magnesium and copper sulphates are plotted in Fig. 6.1. Any possible effects of the temperature difference between the cryoscopic results and the others will be ignored. The values of K are in very satisfactory agreement for values of d between about 6 Å and 9 Å, over which range the variation of K with d is less than 20 per cent. There are theoretical objections to the use of a value of d less

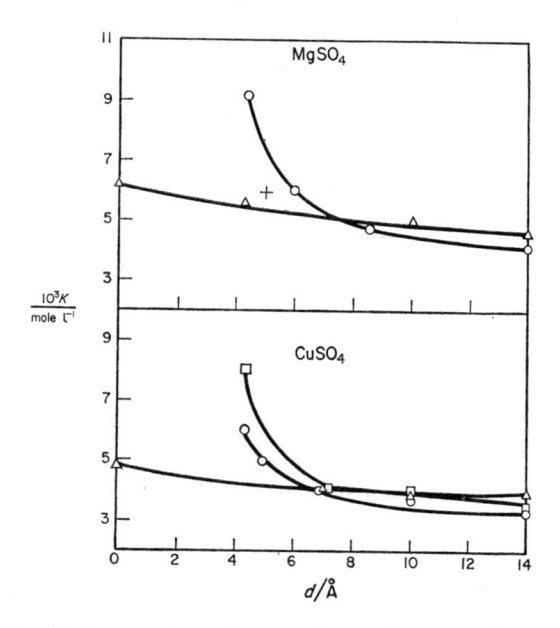

Fig. 6.1. Dissociation constants of magnesium and copper sulphates in water. ○ Cryoscopic (0°C); △ Conductance (25°C); □ Spectrophotometric (25°C); + e.m.f. (mixed electrolyte solution, 25°C).

than about 8 Å in the Debye–Hückel activity coefficient formula for bi-bivalent salts. On the other hand, it was shown earlier (Table 6.1) that for magnesium sulphate the cryoscopic results are slightly better fitted with a d value less than this; this may be because the neglect of the interaction of free ions with associated pairs of ions becomes a better approximation as d decreases.

A standard formula which has often been used is

$$-\log \gamma_i = A' z_i^2 \{I'^{\frac{1}{2}}/(1 + I'^{\frac{1}{2}}) - 0.20\, I'\} \qquad (6.2.5)$$

which corresponds up to $I' = 0.1$ mole kg^{-1} to 6.2.4 with $d \simeq 4.3$ Å; a revision[5] of this formula corresponds to a value of $d \simeq 4.8$ Å. The discrepancies in Fig. 6.1 suggests that these d values are too small for solutions of bi-bivalent salts. A distance of 7 Å would be a reasonable compromise position at which to set the boundary between free and paired ions. In mixed electrolyte solutions, however, d is an average distance which relates to interactions between several kinds of ions. It may then well be that, as in the analysis of the e.m.f. results in 4.4 to obtain the point plotted in Fig. 6.1, a smaller value of d is appropriate.

The cryoscopic method is inevitably restricted to a single temperature and it is not in general suited for measurements on mixed electrolytes. It is possible, however, to make measurements with a solution saturated with a swamping electrolyte at a eutectic or transition point. When the effect of an added solute on the eutectic or transition point is studied, ions common to the swamping salt have no effect on the transition temperature. The method has been used[6] for species identification in the case of polymerized anions ($Na_2SO_4 \cdot 10H_2O/Na_2SO_4$ transition), and Kenttämaa[7] has studied the effect of bi-bivalent sulphates on eutectic points (ice/KNO_3, ice/$KClO_3$, ice/$KClO_4$). It is, however, surprising that under these conditions the dissociation quotient of the complex $M(SO_4)_2^{2-}$ appears to be less than that of MSO_4.

The cryoscopic method has also been used[8] to study equilibria in sulphuric acid. It has a convenient freezing point (10·37°C) and a large cryoscopic constant (6·12 deg mole^{-1} kg) and interionic forces are weaker than in water because of the large dielectric constant (120 at 10°C). The large self-dissociation of the solvent is, however, a complication.

6.3. Isopiestic measurements

The isopiestic technique is a powerful and general method for determining osmotic coefficients and thence activity coefficients of salts from the vapour pressure of the solvent. It is most useful at concentrations above about 0·1 mole l^{-1}, for below this the relative lowering of vapour pressure by the solute is so small that the relative experimental error is high. Incomplete dissociation of an electrolyte results in abnormally low activity coefficients but quantitative conclusions depend on a standard for estimating the activity coefficient γ_i of a completely dissociated electrolyte at ionic strengths of 0·1 mole l^{-1} and above. Rather than use a theoretical equation as a standard it is interesting to equate γ_i with experimental values for a standard electrolyte which is assumed to be completely dissociated. Sodium and calcium chlorides are reasonable choices for uni-univalent and for bi-univalent and uni-bivalent electrolytes respectively. In Table 6.3 activity coefficients γ_i of sodium chloride are compared with some unusually low activity coefficients γ_s of five other salts.[9] Values of the degree of dissociation are calculated from $\alpha = \gamma_s/\gamma_i$. The assumption that specific effects are due to binary interactions alone is a drastic

TABLE 6.3
Activity coefficients in water at 25°C

m mole kg^{-1}	NaCl	KNO$_3$	AgNO$_3$	LiOH	CsI	TlOAc
0·5	0·681	0·545	0·536	0·583	0·599	0·589
1·0	0·657	0·443	0·429	0·523	0·533	0·515
2·0	0·668	0·333	0·316	0·467	0·470	0·444

TABLE 6.4
pK values in water at 25°C

Salt	KNO$_3$	AgNO$_3$	LiOH	CsI	TlOAc
p$K(\gamma)$	0·06	0·10	−0·03	−0·08	−0·11
p$K(\varLambda)$	−0·2	−0·2	−0·08	—	−0·12

one at concentrations above 0·5 molal. However, substituting for α in the equation

$$\frac{(\alpha m \gamma_i)^2}{(1-\alpha)m\gamma_u} = K \qquad (6.3.1)$$

one obtains

$$\frac{\gamma_s^2 m}{1 - (\gamma_s/\gamma_i)} = K\gamma_u \qquad (6.3.2)$$

from which $K\gamma_u$ can be calculated. If log γ_u varies linearly with m p($K\gamma_u$) plotted against m should be a straight line. It will be seen from Fig. 6.2 that this is approximately so; the slopes of the lines are reasonable and the extrapolated pK values given in Table 6.4 are of the same order as those[10] calculated from conductance measurements on solutions one hundred times more dilute. For such small pK values the agreement is reasonable. The numerical values are inevitably very sensitive to the assumptions made to obtain them. The agreement is better for the salt KPF_6 with a smaller dissociation constant for which[11] isopiestic results up to 0·5 mole kg⁻¹ give $pK = 0.32$ (with a standard equation for γ_i) and conductance results up to 0·02 mole kg⁻¹ give $pK = 0.38$ (with the simplified Fuoss–Onsager equation for Λ_i).

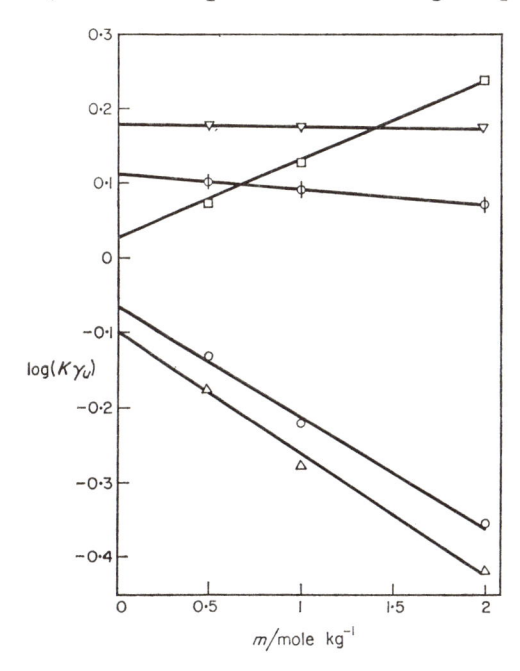

FIG. 6.2. Dissociation constants of uni-univalent salts. ○ KNO_3; △ $AgNO_3$; LiOH; ▽ CsI ϕ TlOAc.

Values of the degree of dissociation α of the associated species in 0·1 mole kg⁻¹ solutions of some higher-valent salts have been calculated from activity coefficient data[9] in a similar way and the results are given in Table 6.5. The formula relating γ_s, γ_i and α is

$$4(\gamma_s/\gamma_i)^3 = \alpha(1 + \alpha)^2 \quad (6.3.3)$$

Calcium chloride was used as a standard for γ_i; a second round of calculations was necessary and γ_i at ionic concentrations below 0·1 mole kg⁻¹ was calculated from a formula[12] which fits the best data for the activity coefficient values of calcium chloride over the whole range 0·003 to 0·3 mole kg⁻¹. To calculate K from $K = \alpha^2 m \gamma_2/(1 - \alpha)$ values of γ_2 were calculated from

$$-\log \gamma_i = z_i^2 \{A'I'^{\frac{1}{2}}/(1 + \rho'I'^{\frac{1}{2}}) - B_i I'\} \quad (6.3.4)$$

with $z_i = 2$, $\rho' = 1.5357$ mole$^{-\frac{1}{2}}$ kg$^{\frac{1}{2}}$ and $B_i = 0.028$ mole^{-1} kg (this equation was chosen because for γ_{CaCl_2} it gives the Guggenheim–Stokes formula[12]). The agreement of the pK values with conductimetric values[10] (Table 6.5) is surprisingly good. In the case of cadmium chloride the assumption that only CdCl$^+$ is present is doubtful. It is interesting that with zinc chloride, although at a molality of 0.1 the value of the activity coefficient of 0.515 is very close to that of calcium chloride, at higher concentrations large differences appear. For example at a

TABLE 6.5

Activity coefficients (0.1 mole kg^{-1}) and pK values in water at 25°C

Salt	CaCl$_2$	Ba(NO$_3$)$_2$	Pb(NO$_3$)$_2$	Mg(OAc)$_2$	CdCl$_2$	K$_2$SO$_4$
γ	0.518	0.431	0.405	0.450	0.228	0.436
a	1	0.713	0.619	0.773	0.170	0.721
p$K(\gamma)$	—	0.92	1.11	0.77	2.08	0.90
p$K(\varLambda)$	—	0.92	1.18	0.78	2.0	0.9

molality of 1 the values are $\gamma_{CaCl_2} = 0.500$, $\gamma_{ZnCl_2} = 0.339$; zinc chloride is no longer completely dissociated.

It is often convenient[13,12] to treat the salts discussed in this section as completely dissociated electrolytes with abnormally large negative specific ionic interaction coefficients (equation 1.3.6).

REFERENCES

1. BROWN, P. G. M. and PRUE, J. E., *Proc. Roy. Soc.*, 1955, **A. 232**, 320.
2. HARNED, H. S. and OWEN, B. B., *The Physical Chemistry of Electrolytic Solutions*, Reinhold, New York, 3rd ed., 1958, p. 176.
3. GUGGENHEIM, E. A., *Trans. Faraday Soc.*, 1960, **56**, 1152.
4. MAYER, J. E., *J. chem. Phys.*, 1950, **18**, 1426.
5. DAVIES, C. W., *Ion Association*, Butterworths, London, 1962, p. 41.
6. TOBIAS, R. S., *J. Inorg. Nuclear Chem.*, 1961, **19**, 348.
7. KENTTÄMAA, J., *Acta chem. Scand.*, 1958, **12**, 1323.
8. BASS, S. J., GILLESPIE, R. J. and ROBINSON, E. A., *J. chem. Soc.*, 1960, 821.
9. ROBINSON, R. A. and STOKES, R. H., *Electrolyte Solutions*, Butterworths, London, 1959, pp. 491–502.
10. DAVIES, C. W., *The Structure of Electrolytic Solutions*, Ed. HAMER, W. J., John Wiley, New York, 1959, p. 19.
11. ROBINSON, R. A., STOKES, J. M. and STOKES, R. H., *J. phys. Chem.*, 1961, **65**, 542.
12. GUGGENHEIM, E. A. and STOKES, R. H., *Trans. Faraday Soc.*, 1958, **54**, 1646.
13. GUGGENHEIM, E. A. and TURGEON, J. C., *Trans. Faraday Soc.*, 1955, **51**, 747.

CHAPTER 7

RATES OF REACTION

7.1. Introduction

One of the earliest methods of assessing the " strength " of an acid was from its catalytic power on reactions such as the hydrolysis of esters or the inversion of sucrose. In suitable cases measurements of reaction rates are capable of giving values of acidity constants of good precision. The rate constant for a second-order reaction between two species A and B is defined by $-d[A]/dt = k[A][B]$. If k and the concentration of A or B in a solution are known, the other concentration can be found by a rate measurement. Unfortunately, rate constants are not in general independent of the environment, the influence of which is given by $k = k^0 f_A f_B / f_x$ where f_A and f_B are activity coefficients of the reactants and f_x that of the activitated complex (it is usual to define the activity coefficients as unity in the pure solvent). This environmental effect in electrolyte solutions is known as the primary salt effect, and it is clear that its effect must be known or calculable before rate measurements can be used to measure a concentration. If the activity coefficients are given by equation 1.3.4 one obtains

$$\log k = \log k^0 + \frac{2A z_A z_B I^{\frac{1}{2}}}{1 + I^{\frac{1}{2}}} + BI \qquad (7.1.1)$$

where B is an empirical parameter. This equation successfully fits[1] the data for some reactions in solutions of strong electrolytes up to an ionic strength of about 0·1 mole l⁻¹, but there are other cases[2] where even in dilute solution the specific interactions are too strong to be adequately taken account of by the parameter B. Complex formation involving a reactant will have a negative effect on the rate, whilst if it involves the activated complex the reaction will be catalysed. If complex formation is weak its influence can be formally described as an effect on the activity coefficient factor rather than as a change in the concentration of a reactant or the contribution of a new reaction. An equilibrium involving a reactant can be satisfactorily studied only if other specific effects are negligible or can be allowed for.

7.2. Protolytic equilibria

The rate constants of reactions which have first order kinetics are particularly easy to measure accurately. If a species A catalyses the reaction of a substrate B the value k[A] is obtained from the slope of the first order plot. k is often known as the catalytic coefficient of A. If the substrate is a neutral molecule $z_B = 0$ and the second term on the right-hand side of 7.1.1 disappears. The hydrolysis of dimethylacetal in water is specifically catalysed by hydrogen ions and measurements at 20°C with dilute solutions of the completely dissociated perchloric acid over the concentration range 1×10^{-3} to 5×10^{-3} mole l^{-1} give[3] an almost constant value $k/[\mathrm{H}^+] = 9\cdot03$ l mole^{-1} min^{-1}. This

TABLE 7.1
pK of chloroacetic acid

$10^3 a$/mole l^{-1}	3·00	5·00	10·00	15·0
$10^3 [\mathrm{H}^+]$/mole l^{-1}	1·51	2·12	3·26	4·16
$10^3 Q$/mole l^{-1}	1·53	1·55	1·58	1·60
$10^3 K$/mole l^{-1}	1·40	1·40	1·39	1·39

implies that B in equation 7.1.1 is zero for the solutions used, and if it is assumed that B is also equal to zero in dilute solutions of chloroacetic acid of concentration a, the value of $k/[\mathrm{H}^+]$ can be used to calculate the hydrogen ion concentrations in dilute solutions of the acid. This is a reasonable assumption provided that the range of a is such that the ionic concentrations are similar to those in the perchloric acid solutions. The values of $[\mathrm{H}^+]$ in Table 7.1 are obtained and lead to the values tabulated of the equilibrium quotient and acidity constant (B in equation 1.3.5 is set equal to zero in calculating ionic activity coefficients). The acidity constant values are constant and the mean value of $1\cdot39_5 \times 10^{-3}$ mole l^{-1} is in excellent agreement with a value[4] of $1\cdot39_6 \times 10^{-3}$ mole l^{-1} obtained from precise conductivity measurements.

Accurate measurements have been made[5] at 25°C on the rate of inversion of sucrose in sodium bisulphate–sodium sulphate buffers. If the value of B in 7.1.1 is set equal to the value found from measurements in HCl, HBr, HClO$_4$ and HNO$_3$, the kinetic results are consistent with a pK of 1·99. The same value is obtained from e.m.f. measurements if a reasonable but arbitrary assumption is made about the ion-size term used to calculate $\gamma_H\gamma_{Cl}$ in order to convert values of $[H^+]\,\gamma_H\gamma_{Cl}$ to $[H^+]$ (see 4.3).

It will be obvious that general acid or base-catalysed reactions are much less suitable for the study of protolytic equilibria.

7.3. Complex ion equilibria

Measurements of the rate of a specifically hydroxyl-ion catalysed reaction, the depolymerization of diacetone alcohol, have been used[6] to study the stability of hydroxo-complexes of cations (in other words, the acidity constants of the hydrated cations). The value of B in equation 7.1.1 is small for sodium, potassium and rubidium hydroxides and certainly negligible up to an ionic strength of 0·1 mole l⁻¹. Yet

TABLE 7.2

pK_d of CaOH⁺

$2 \times 10^2\,[Ca(OH)_2]$/mole l⁻¹	1·442	1·802	2·162	2·523	2·882	3·243	3·569
$10^2[OH^-]$/mole l⁻¹	1·347	1·667	2·002	2·281	2·606	2·874	3·180
$10^2 K_d$/mole l⁻¹	5·0	5·0	5·9	4·6	5·1	4·4	5·0

in the case of calcium hydroxide solutions, for example, k has the value 0·447 l mole⁻¹ min⁻¹ in 0·035M solution compared with that of 0·502 in the alkali metal hydroxides. The decrease is reasonably ascribed to the formation of CaOH⁺ and if B is assumed to have the same value (zero) that it has in the alkali metal hydroxides the values of $[OH^-]$ given in Table 7.2 are obtained. These lead to the values shown in the final row for the dissociation constant of CaOH⁺ (B in equation 1.3.5 is again set equal to zero). Thus $pK_d = 1·29$. The same technique had been used[5-8] to obtain dissociation constants for BaOH⁺, TlOH, $(CH_3)_2$TlOH and $Co(NH_3)_6OH^{2+}$ with $pK_d = 0·64, 0·49, 1·04$ and 1·85 respectively. The first two values, however, are small and depend on results at concentrations at which the validity of the assumption that $B = 0$ in 7.1.1 is doubtful. In tetraalkylammonium hydroxide solutions[9] k increases with concentration about as much as it decreases in the cases of barium and thallous hydroxides.

A value $pK_d = 1·34$ was obtained[10] for CaOH⁺ at 0°C by comparing the rates of hydrolysis of the ester $^+NEt_3 \cdot CH_2 \cdot CO_2Et$ in alkaline solutions of calcium chloride and sodium chloride at the same ionic strength.

Added calcium chloride, however, had no effect[10] on the rate of hydrolysis of ethyl acetate in sodium hydroxide. The effect of formation of CaOH+ on adding calcium ions is offset by an increase in $f_A f_B/f_x$. Presumably the activated complex sufficiently resembles a hydroxyl ion for calcium ions to form complexes with it also; in other words an additional reaction occurs. For a similar reason dissociation constants for calcium, barium and zinc complexes with carboxylate anions, obtained[11] from measurements of the anion concentration by its effect on the general base-catalysed decomposition of nitramide, are higher than those obtained by other methods.

An example[12] of a different type of reaction where complex formation with both a reactant ion and the activated complex occurs is the reaction between bromoacetate and thiosulphate ions in the presence of doubly charged cations. On the other hand the retardation of the reaction between ferric and iodide ions by hydroxyl and sulphate ions[13] is quantitatively consistent with the supposition that the activated complex of composition FeI_2^+, which only carries a single positive charge, does not form complexes with the anions. It will be clear that care is necessary in obtaining dissociation constants from rate measurements.

It is appropriate to mention finally that the rate of a reaction at a dropping mercury electrode has been used in a few cases as a measure of the concentration of a species in solution. In 4.7 it was seen that the greatest possible current (the diffusion current i_D) at a dropping mercury electrode is given by $i_D = k_M[M^{x+}]$. This current can itself be used as a measure of the concentration of free M^{x+} in a solution provided k_M is known and the rate of dissociation of the complex into ligand and free cation is sufficiently slow compared with the time each mercury drop remains at the electrode. The conditions for the applicability of this method (amperometry) are therefore the opposite of those necessary for the validity of the polarographic method discussed in 4.7. The reduction of the complex itself may, of course, give a second polarographic wave the i_D for which can be likewise used as a measure of the concentration of the complex. In this way equilibria involving ethylenediaminetetra-acetate complexes such as

$$CuY^{2-} + Ga^{3+} \rightleftharpoons Cu^{2+} + GaY^-$$

have been studied[14] by measurements of the diffusion currents for CuY^{2-} and Cu^{2+}.

REFERENCES

1. GUGGENHEIM, E. A. and PRUE, J. E., *Physicochemical Calculations*, North-Holland, Amsterdam, 1956, p. 466.
2. INDELLI, A. and PRUE, J. E., *J. chem. Soc.*, 1959, 107.
3. GUGGENHEIM, E. A. and PRUE, J. E., *Physicochemical Calculations*, North-Holland, Amsterdam, 1956, p. 477.
4. IVES, D. J. G. and PRYOR, J. H., *J. chem. Soc.*, 1955, 2104.
5. GUGGENHEIM, E. A., HOPE, D. A. L. and PRUE, J. E., *Trans. Faraday Soc.*, 1955, **51**, 1387.
6. BELL, R. P. and PRUE, J. E., *J. chem. Soc.*, 1949, 362.
7. LAWRENCE, J. K. and PRUE, J. E., " Int. Conf. Co-ordination Chem.", *Chem. Soc. Special Publ.*, 1959, No. 13, p. 186.
8. CATON, J. A. and PRUE, J. E., *J. chem. Soc.*, 1956, 671.
9. HALBERSTADT, E. S. and PRUE, J. E., *J. chem. Soc.*, 1952, 2234.
10. BELL, R. P. and WAIND, G. M., *J. chem. Soc.*, 1950, 1979.
11. BELL, R. P. and WAIND, G. M., *J. chem. Soc.*, 1951, 2357.
12. WYATT, P. H. and DAVIES, C. W., *Trans. Faraday Soc.*, 1949, **45**, 774.
13. SYKES, K. W., *J. chem. Soc.*, 1952, 124.
14. SCHWARZENBACH, G., GUT, R. and ANDEREGG, G., *Helv. chim. acta*, 1954, **37**, 937.

CHAPTER 8

RELAXATION SPECTROMETRY

8.1. Introduction

In the experiments discussed in the last chapter, the rate of a reaction of known rate constant was used as a measure of the equilibrium concentration of a species involved in an equilibrium. In this chapter it will be seen that a kinetic technique can sometimes yield equilibrium constants which are not otherwise susceptible of measurement. Relaxation spectrometry, which has developed as a physico-chemical technique largely as the outcome of work in Göttingen,[1] has strikingly extended the range of reaction half-times accessible to experimental study. Flow techniques can be used for the measurement of rates of reaction with half-times down to 10^{-3} sec (the limiting factor is the speed of mixing the reactants), and relaxation spectrometry fills the gap between 10^{-3} sec and the half-time for a bimolecular diffusion-controlled reaction, which is about 10^{-9} sec for two singly and oppositely charged ions in 0·01M solution in water. The rates of many ionic reactions fall in this region and their study is important not only for its own sake but also for another reason. The ratios of some of the rate constants correspond to the equilibrium constants for the distribution between states indistinguishable by ordinary equilibrium measurements. This is particularly true of equilibria involving solvent molecules.

8.2. Principles of method

In conventional kinetic studies the reactants are mixed to give a system far removed from equilibrium and the rate of the subsequent reaction is followed by physical or chemical methods. In relaxation kinetics one of the physical conditions (e.g. pressure or temperature) of a system in equilibrium is suddenly changed slightly and the relaxation of the system towards a new position of equilibrium is followed. Of course the physical condition which is changed (the "forcing parameter") must be one which causes a change in the equilibrium constant of the reaction being investigated. If the difference in concentration

Δc between the concentration c of a reactant and the equilibrium concentration $\bar{c}$ which it will eventually reach is sufficiently small then

$$dc/dt = -\Delta c/\tau \qquad (8.2.1)$$

where the constant τ is the relaxation time of the system. In words, the rate of change of c is assumed to be directly proportional to its difference from the equilibrium concentration $\bar{c}$. The equation is assumed to be valid for any elementary reaction whatever its order.

As an example, consider the following elementary reaction

$$A + B \underset{k_2}{\overset{k_1}{\rightleftharpoons}} AB$$

for which

$$-\frac{d[A]}{dt} = k_1[A][B] - k_2[AB] \qquad (8.2.2)$$

The combination of 8.2.1 and 8.2.2, setting $dc/dt = d[A]/dt$ and $\Delta c = [A] - \overline{[A]} = [B] - \overline{[B]} = \overline{[AB]} - [AB]$, gives

$$\Delta c/\tau = k_1(\Delta c + \overline{[A]})(\Delta c + \overline{[B]}) - k_2(\overline{[AB]} - \Delta c) \qquad (8.2.3)$$

which, with $\overline{[AB]}/\overline{[A][B]} = k_1/k_2$ and neglecting the term $k_1(\Delta c)^2$ gives

$$\tau^{-1} = k_2 + k_1(\overline{[A]} + \overline{[B]}) \qquad (8.2.4)$$

Measurements of the relaxation time as a function of the concentration of A and B therefore permits evaluation of the rate constants. Similarly, for a reversible first-order isomerization equilibrium $\tau^{-1} = k_1 + k_2$.

A reaction made up of a succession of elementary reactions gives a " spectrum " of relaxation times (whence the term relaxation spectrometry). In general each relaxation time is a function of several rate constants and not of those for one of the elementary reactions alone. There remains, of course, the usual ambiguity associated with reaction kinetics; that a set of rate constants consistent with a particular mechanism fit the experimental data does not prove that another set of rate constants associated with a different mechanism could not be found capable of giving as good or an even better fit of the results. In the single perturbation method, the relaxation times are found by analysis of Δc versus time curves, just as c–t curves are analysed to give the rate constants themselves in conventional kinetics. There are, however, a group of methods which give the relaxation times directly. In these the perturbation is applied periodically to the system. For example, an ultrasonic wave passing through a solution is a periodically applied pressure perturbation (the associated temperature wave is in the case of water near room temperature negligible; it is zero at 4°C, the temperature of maximum density). If the relaxation time of an equilibrium is much less than the period of the pressure wave the position of equilibrium will follow the pressure fluctuations and keep in phase with them. If, however, the relaxation time is of the same order as the period of the pressure wave, the position of equilibrium will lag behind; there will be a phase displacement between change of pressure and the position of equilibrium, that is between change of pressure and the local density. The maintenance of the system in a state of disequilibrium requires absorption of sound energy. Quantitatively the excess absorption over that of pure water is given by

$$I = I_0 e^{-Qcl} \qquad (8.2.5)$$

with

$$Q = \frac{2\pi}{\lambda c} \cdot \frac{\kappa^{Ch}}{\kappa^0} \cdot \frac{2\pi\nu\tau}{1 + (2\pi\nu\tau)^2} \qquad (8.2.6)$$

I_0 and I are the respective intensities of the incident and emergent sound beam of wavelength λ (frequency ν) passing through a solution of concentration c and length l (note the analogy between 8.2.5 and the Lambert–Beer Law). κ^{Ch} is the contribution to the compressibility of the solution from the mobile chemical equilibrium, and κ^0 is the remaining part. For dilute aqueous solutions, κ^0 can be equated to the compressibility of water. κ^{Ch} is given by

$$\kappa^{Ch} = -\frac{1}{V}\left(\frac{\partial V}{\partial \alpha}\right)_p \frac{d\alpha}{dp} = \frac{c\alpha(1-\alpha)}{(2-\alpha)}\frac{(\Delta V)^2}{RT} \qquad (8.2.7)$$

where α is a degree of dissociation at concentration c and ΔV the molar volume change associated with the reaction. By measurement of Q as a function of ν, the relaxation time is easily obtained because the frequency factor $2\pi\nu\tau/\{1 + (2\pi\nu\tau)^2\}$ has a maximum at $2\pi\nu = \tau^{-1}$. From absolute values of Q, κ^{Ch} and thence α is also obtainable provided ΔV is known or can be estimated.

8.3. Experimental techniques

As single perturbation forcing parameters for ionic equilibria, changes of pressure, temperature and electric field have been used.

The use of the last depends on the increase in the dissociation constant of an electrolyte at high field strengths, sometimes known as the second Wien effect. The effect has also been used directly[2] as a method of measuring the dissociation constants of species such as lanthanum ferricyanide and the bi-bivalent sulphates.

In one application[3] of the pressure-jump method a pressure of about fifty atmospheres on the solution is suddenly released by puncturing a phosphor-bronze membrane with a steel needle. The change of conductance of the solution as it relaxes to equilibrium is followed by applying the out-of-balance bridge current to an oscilloscope. The opposite arm of the bridge contains a solution which has no chemical relaxation effect in the region accessible with the p-jump method, of concentration adjusted to balance the bridge when all the solutions are at atmospheric pressure. The 50,000 c/sec current used also provides a time scale. The pressure fall is not an instantaneous one and is characterized by a relaxation time of the order of 0.5×10^{-4} sec. This limits the method to the study of reactions of half-times between 0.5×10^{-4} sec and 50 sec.

A rise of about 10°C in the temperature of a solution can be achieved[4] in a few microseconds by the discharge of a condenser raised to a potential of about 20 kV. The reactions have been followed spectrophotometrically. The method is obviously one of general applicability and refinements seem likely to extend its useful range to half-times as small as 10^{-7} sec. Even if the reaction of interest has $\Delta H \sim 0$, it can sometimes be coupled to a reaction which is temperature sensitive.

The field dissociation method is also applicable for reactions of half-times down to 10^{-7} sec. For half-times between 10^{-7} and the theoretical limit of 10^{-9} sec the sound absorption method[5] is pre-eminent. Various techniques are available.[6]

8.4. Results

One of the simplest applications of relaxation spectrometry is in the study of the protolytic reactions of acids and bases, to which equation 8.2.4 is applicable. It is striking that in reactions involving H^+ or OH^- in water the second-order rate constants, even with weak acids and bases, are often[6] close to the diffusion-controlled theoretical limit (*ca.* 5×10^{10} l mole^{-1} sec^{-1}). Exceptions are when C–H links are involved or when the proton to be detached is involved in an intramolecular hydrogen bond. Another exception is with dilute sulphurous acid solutions.[7] The occurrence of sound absorption at unusually low frequencies (10^6 to 10^7 c/sec) and the magnitude of the absorption coefficients leads to the conclusion that the responsible reaction is not the simple protonation but one that involves simultaneous dehydration, viz. $H^+ + HSO_3^- \rightleftarrows SO_2 + H_2O$ with $k_1 = 2 \times 10^8$ l mole^{-1} sec^{-1} and $k_2 = 3 \cdot 4 \times 10^6$ sec^{-1} (at an ionic strength of about 0·1 mole l^{-1}). Provided that the identification of the reaction is correct, the closeness of the equilibrium constant $K'_a = k_2/k_1 = 1 \cdot 7 \times 10^{-2}$ mole l^{-1} to the conventional acidity constant of $1 \cdot 72 \times 10^{-2}$ mole l^{-1} at 25°C shows that very little of the sulphur dioxide is present in the hydrated form H_2SO_3, which agrees with the conclusions of spectroscopic work.

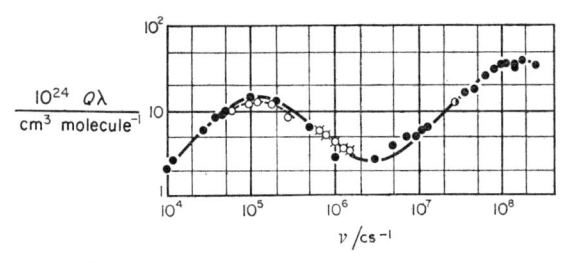

FIG. 8.1. Sound absorption of aqueous magnesium sulphate solutions (0·01 to 0·1M). (From reference 5.)

This is one illustration of the potentiality of relaxation spectrometry in distinguishing between states not distinguishable by equilibrium measurements alone.

Of particular interest in the present context are sound absorption measurements on bi-bivalent sulphates.[5] They show two absorption maxima which do not occur if either cation or anion is replaced by a singly-charged ion. The absorption curve for magnesium sulphate in Fig. 8.1 is a typical example. The high frequency maxima are around 10^8 c/sec and do not depend on the cation (at even higher frequencies there is an absorption characteristic of all electrolytes and probably due to the relaxation of the ion atmospheres). The position of the low frequency maxima is specifically dependent on the cation and varies over several powers of ten (see Table 8.1). A low frequency maximum is difficult to detect with copper sulphate and absent with zinc sulphate.

The simplest interpretation of the two absorption maxima is to ascribe them to two stages of complex formation. In the first stage, outer-sphere complex formation, the associated ions are still separated by a co-ordination sphere of water molecules attached to the cation.

The rate constants and consequently the τ^{-1} values for this process should be high and not specifically dependent on the cation. This explains the high frequency maxima. For inner-sphere complex formation in which the ions come into direct contact, one expects the rate constants and the associated τ^{-1} values to be lower and specifically dependent on the cations. This is what is observed. If the degree of dissociation into free ions were negligible the equilibrium between outer- and inner-sphere complexes would correspond to a simple reversible first-order isomerization reaction for which τ^{-1} is equal to

TABLE 8.1
Sound absorption in bi-bivalent sulphates

Cation	Be^{2+}	Mg^{2+}	Mn^{2+}	Co^{2+}	Ni^{2+}
$10^{24}(Q\lambda)_{\max}/\text{cm}^3$ molecule^{-1}	5	15	50	20	20
$10^{-6}\tau^{-1}/\text{sec}^{-1}$	0·006	0·76	19	2·5	0·06

TABLE 8.2
Rate constants for outer-sphere $\rightleftharpoons$ inner-sphere conversion in bi-bivalent sulphates

Cation	k_{34}/sec^{-1}	k_{43}/sec^{-1}
Be^{2+}	1×10^2	$1\cdot5 \times 10^3$
Mg^{2+}	1×10^5	8×10^5
Mn^{2+}	4×10^6	2×10^7
Fe^{2+}	1×10^6	6×10^6
Co^{2+}	2×10^5	$2\cdot5 \times 10^6$
Ni^{2+}	$1\cdot5 \times 10^4$	1×10^5
Zn^{2+}	3×10^7	$>10^8$

the sum of the rate constants for the forward and backward reactions. The degree of dissociation is, of course, not negligible and Eigen and Tamm[5] conclude from a detailed analysis of results for magnesium sulphate over a range of concentration that a model with three discrete stages of association gives the best account of them consistent with what is known independently from equilibrium data about the overall association constants and ΔV values. They write

$$\text{M}^{2+} + \text{SO}_4^{2-} \underset{k_{21}}{\overset{k_{12}}{\rightleftharpoons}} \text{M}^{2+}(\text{H}_2\text{O})_2\text{SO}_4^{2-} \underset{k_{32}}{\overset{k_{23}}{\rightleftharpoons}} \text{M}^{2+}(\text{H}_2\text{O})\text{SO}_4^{2-} \underset{k_{43}}{\overset{k_{34}}{\rightleftharpoons}} \text{M}^{2+}\text{SO}_4^{2-}$$

where water molecules in the co-ordination sphere of the cation are omitted except for those interposed directly between the ions. The rate constants k_{12}, k_{21}, k_{23}, k_{32} have the same value for all sulphates and $k_{12} = 4 \times 10^{10}$ l mole^{-1} sec^{-1}, $k_{21} = k_{23} = k_{32} = 10^9$ sec^{-1}. The rate constants for the final stage are given in Table 8.2. One notes that k_{43} is the largest in all cases and close to the τ^{-1} values in Table 8.1. In the detailed analysis k_{34} was obtained by combining k_{43} with values for the equilibrium constant k_{43}/k_{34} obtained from the values of $(Q\lambda)_{max}$ (see equations 8.2.6 and 8.2.7). In spite of specific differences in the rate constants, the equilibrium constant k_{43}/k_{34} which corresponds to the relative concentrations of outer and inner-sphere complexes, is approximately ten in all cases. In other words, only about 10 per cent of associated ions are in direct contact. This is probably why the thermodynamic properties of solutions of the bi-bivalent sulphates are so similar. The smallness of the low frequency absorption for copper sulphate is probably due to the square-planar bound water molecules being very firmly attached, whilst sulphate ions which intrude into the axial positions do so as easily as they can enter the outer sphere, and therefore count as being in the outer-sphere population.

The rate constants themselves suggest that the contact interactions between the cations and water or the anions show large specific differences. The difference between the constants for nickel and magnesium which have identical ionic radii is striking. It is predicted by ligand field theory. The same k_{34} value as that for magnesium sulphate is found not only for the chromate and thiosulphate but also by T-jump studies[8] for the very stable magnesium complexes of chelating agents (e.g., edta). This indicates that a water molecule has to be completely removed before a ligand attaches itself; the reaction is in physical organic chemists' terminology $S_N 1$. Extensive studies of rates of complex formation[9] show that this behaviour is general when rate constants are less than 10^7 sec^{-1}, but not so low that deprotonation of the aquated cation becomes an intermediate step in inner-sphere complex formation, as with, for example, beryllium.

The rate constants for the beryllium ion are low enough for the corresponding relaxation processes to have a half-time accessible to study by the single-perturbation p-jump method.[3] The same is true[10,11] for the formation of the complexes $AlSO_4^+$ and $FeCl^{+2}$. The relative populations in the outer- and inner-sphere states are 10^{-1} to 10^{-2} and about 1 in the two cases. The rate constants are even less for the reaction $Co(NH_3)_5H_2O^{3+} SO_4^{2-} \rightleftharpoons Co(NH_3)_5SO_4^+ + H_2O$, which can be followed by conventional spectrophotometry.[12] In the case of chromic salts the reaction takes weeks at room temperature.

It is clear that relaxation spectrometry provides an important method of obtaining new information about the distribution of ions in solution, even if in its present stage of development the numerical values of the rate and equilibrium constants are less certain than those obtainable by classical techniques.

REFERENCES

1. EIGEN, M., *Z. Elektrochem.*, 1960, **64**, 115 ; DE MAEYER, L., *Z. Elektrochem.*, 1960, **64**, 65 ; TAMM, K., *Z. Elektrochem.*, 1960, **64**, 73.
2. PATTERSON, A. and FREITAG, H., *J. Electrochem Soc.*, 1961, **108**, 529.
3. STREHLOW, H. and WENDT, H., *Inorg. Chem.*, 1963, **2**, 6.
4. CZERLINSKI, G. and EIGEN, M., *Z. Elektrochem.*, 1959, **63**, 65.
5. EIGEN, M. and TAMM, K., *Z. Elektrochem.*, 1962, **66**, 93, 107 ; TAMM, K., *Handbuch der Physik*, Vol. XI/I, Acoustics I, Ed. FLUGGE, S., Springer-Verlag, Berlin, 1961, p. 202.
6. EIGEN, M. and DE MAEYER, L., *Technique of Organic Chemistry*, Vol. VIII, Ed. FRIESS, S. L., LEWIS E. S. and WEISSBERGER, A., 2nd ed., Interscience, New York, 1963, p. 895.
7. EIGEN, M., KUSTIN, K. and MAASS, G., *Z. phys. Chem. (Frankfurt)*, 1961, **30**, 130.
8. DIEBLER, H. and EIGEN, M., *Z. phys. Chem. (Frankfurt)*, 1959, **20**, 299.
9. EIGEN, M., *Pure and applied Chem.*, 1963, **6**, 97.
10. BEHR, B. and WENDT, H., *Z. Elektrochem.*, 1962, **66**, 223.
11. WENDT, H. and STREHLOW, H., *Z. Elektrochem.*, 1962, **66**, 226.
12. TAUBE, H. and POSEY, F. A., *J. Amer. chem. Soc.*, 1953, **75**, 259.

CHAPTER 9

ACIDITY CONSTANTS

9.1. Introduction

A massive quantity of information on protolytic equilibria is available. Not only have acidity constants been determined, often with high accuracy, but also in many instances their temperature dependence. Our understanding of the nature of the chemical bond is as yet far too inadequate for it to be possible to calculate even an approximate value for the equilibrium constant of an ionic reaction in solution from the fundamental properties of the molecules. Nevertheless the results do show some systematic patterns which can be related to simple theoretical models. In the case of acidity constants this is particularly so for substituent effects on the strengths of organic acids and also for the relative strengths of oxyacids.

Before discussing these topics, however, there are some points of general theory which it is important to be clear about in both this and the next chapter.

For a reaction $A + B \rightleftharpoons C$ the association constant K_{ass} is given by

$$K_{ass} = \frac{(Q_C/NV)}{(Q_A/NV)(Q_B/NV)} \, e^{-\Delta E_0/RT} \qquad (9.1.1)$$

where

$$\Delta E_0 = N(\epsilon_{0C} - \epsilon_{0A} - \epsilon_{0B}) \qquad (9.1.2)$$

is the difference in energy between a mole of C in its lowest energy level and a mole each of A and B also in their lowest levels, and Q_A, Q_B, Q_C are partition functions defined by

$$Q = \Sigma_i \, g_i \, e^{-(\epsilon_i - \epsilon_0)/kT} \qquad (9.1.3)$$

where ϵ_i and g_i refer to the energies and multiplicities of the various states of the molecules. The value of a partition function depends upon the spacing of the energy levels above the lowest level and their multiplicities. Q is increased by closer spacing of the levels or by larger multiplicities. The energy levels for a molecular vibration are fairly far apart, considerably closer for rotation and effectively continuous for translation. For a linear molecule in an ideal gas, Q is given by the standard formula

$$Q = Q_t \times Q_r \times Q_v$$
$$= \left(\frac{2\pi mkT}{h^2}\right)^{3/2} V \times \left(\frac{8\pi^2 IkT}{\sigma h^2}\right) \times \Pi_\nu (1 - e^{-h\nu/kT})^{-1} \quad (9.1.4)$$

where the subscripts t, r and v correspond to the various kinds of motion, m is the mass of the molecule, I its moment of inertia, V is the volume in which it is confined, σ a symmetry number, and the ν's are fundamental vibration frequencies. For molecules in solution the situation is much more complicated. However, for our purposes it suffices to assume that the orders of magnitude of values of f, the partition functions per degree of freedom ($Q_t = f_t^3$ etc.), can be obtained from the same formulae. The values are given in Table 9.1, and it is seen that $f_t \gg f_r > f_v$.

It is clear from equation 9.1.1 that the equilibrium constant of a reaction can be regarded as dependent fundamentally on two factors. The first is ΔE_0 and it is this quantity or changes in it which theories about the energies of molecules (e.g. resonance effects) attempt to predict. The second factor, the partition function quotient, is also important. For example, in the gaseous phase a high association constant for a reaction in which three translational degrees of freedom are converted into one degree of vibrational and two degrees of rotational motion would, because $f_t \gg f_r > f_v$, be disfavoured by the partition function quotient. There is a complication if the reaction occurs in solution. Suppose A and B are oppositely charged; as a consequence when they associate to form uncharged C the process will be accompanied by considerable desolvation. The reaction will now become $AS_m^+ + BS_n^- \rightleftharpoons C + (m+n)S$ where S is a solvent molecule. Released solvent molecules will exchange vibrational motion for translation and rotation and in consequence of this the partition function quotient will now be more favourable to association.

It is unfortunate that neither ΔE_0 nor the partition function quotient are easily related to experimental quantities. Let us consider their relation to the thermodynamic functions $\Delta G^\ominus$, $\Delta H^\ominus$ and $\Delta S^\ominus$ given by the following equations (most of the data in the literature for $\Delta H^\ominus$ and $\Delta S^\ominus$ come from measurements of acidity constants over a temperature

G

range; precise values of acidity constants are required to give even moderately reliable values of $\Delta H^\ominus$ and $\Delta S^\ominus$):

$$\Delta G^\ominus = -RT \ln K_{ass} \qquad \Delta S^\ominus = \frac{d}{dT}(RT \ln K_{ass})$$

$$\Delta H^\ominus = RT^2 \frac{d \ln K_{ass}}{dT} \tag{9.1.5}$$

$$\Delta G^\ominus = \Delta H^\ominus - T\Delta S^\ominus \tag{9.1.6}$$

$$\Delta G^\ominus = \Delta E_0 - RT \ln \{(Q_C/NV)/(Q_A/NV)(Q_B/NV)\} \tag{9.1.7}$$

$$\Delta H^\ominus = \Delta E_0 + RT^2 \, d \ln (Q_C/Q_A Q_B)/dT \tag{9.1.8}$$

$$\Delta S^\ominus = (\Delta H^\ominus - \Delta G^\ominus)/T = R \ln (Q_C/NV)/(Q_A/NV)(Q_B/NV)$$
$$+ RT \, d \ln (Q_C/Q_A Q_B)/dT \tag{9.1.9}$$

TABLE 9.1
Partition functions and their temperature dependence (ca. 300°K)

Motion	f	$R \ln f$ cal deg^{-1} mole^{-1}	RT(d ln f/dT) cal deg^{-1} mole^{-1}	RT^2(d ln f/dT) cal mole^{-1}
Translation	10^8 cm^{-1}	37	1	300
Rotation	10^2	9·2	1	300
Vibration	1	0	0	0

Only in the special case of no temperature dependence in the Q's is $\Delta H^\ominus = \Delta E_0$ and $\Delta S^\ominus$ simply proportional to the logarithm of the partition function quotient. The crucial question is how important the partition function temperature dependence term is in other cases. It is seen from equation 9.1.4 that the partition function for one degree of translational or rotational freedom is proportional to $T^{\frac{1}{2}}$. For one degree of vibrational freedom, f_v will be nearly independent of T unless the vibration frequency is small. From the temperature dependence of the partition functions one obtains the values in the last two columns of Table 9.1. If values from the last column are inserted in equation 9.1.8 it is seen that unless a reaction involves the conversion of several degrees of vibrational freedom to translations and rotations (or viceversa) it is likely that $\Delta H^\ominus$ will be equal to ΔE_0 within a kilocalorie or so. Often, however, it is small changes in ΔE_0 due to a substituent that are under discussion; for this situation it has been argued[1] that $\delta \Delta G^\ominus$ is closer to $\delta \Delta E_0$ than is $\delta \Delta H^\ominus$. Concerning the interpretation of $\Delta S^\ominus$ values, if the values in Table 9.1 are inserted in equation 9.1.9 it will be clear that the first term on the right-hand side is dominant, and so reasoning about the effect on the partition function quotient of conversion of translational and rotational motion into vibrational will apply also to the $\Delta S^\ominus$ values. As an example, $\Delta S^\ominus$ values[2] of 43 cal deg^{-1} mole^{-1} for the reaction $H^+ + PO_4^{3-} \rightarrow HPO_4^{2-}$ and 22·1 cal deg^{-1} mole^{-1} for the reaction $H^+ + OAc^- \rightarrow HOAc$ clearly imply that the former is accompanied by more desolvation. There are a number of discussions in the literature of $\Delta S^\ominus$ values of protolytic equilibria in relation to the solvation and modes of motion of the species involved, but the conclusions are as yet insufficiently firmly established to warrant detailed discussion here. Again, the argument often involves small changes $\delta\Delta S^\ominus$ due to a substituent; in this situation changes in the temperature dependence of partition functions may be comparable in importance with changes in the partition functions themselves.

9.2. Substituent effects on protolytic equilibria

Bjerrum[3] was the first to discuss the change of the ratio of the first and second acidity constants in the series $HO_2C \cdot (CH_2)_n \cdot CO_2H$ in terms of a simple electrostatic model. The two equilibria are

$$HO_2C \cdot (CH_2)_n \cdot CO_2H \rightleftarrows {}^-O_2C \cdot (CH_2)_n \cdot CO_2H + H^+$$

$$^-O_2C \cdot (CH_2)_n \cdot CO_2H \rightleftarrows {}^-O_2C \cdot (CH_2)_n \cdot CO_2^- + H^+$$

It is postulated that the amount of work which has to be done to remove a proton from an acid anion exceeds that required to remove a proton from a neutral molecule by the electrostatic term $e^2/\epsilon r$ where e is the electronic charge, ϵ the dielectric constant, and r the distance between the proton and the charge on the acid anion. This quantity is a free energy and the ratio of the two acidity constants will therefore be $K_1/K_2 = 4 \exp(e^2/\epsilon r kT)$. The factor 4 is inserted because the acidity constants as ordinarily defined take no account of the fact that the neutral molecule can lose a proton at one of two positions and recover it at only one, whilst the opposite is true of the acid anion. Unfortunately the value of r in solution is unknown. It will be approximately 4 Å in $HC_2O_4^-$ depending on the configuration of the molecule and the charge distribution in the carboxylate group and it is reasonable to assume that it increases by 1·26 Å per methylene group, this being the length by which a straight carbon chain would increase along its axis. In Table 9.2 experimental values[4] of ΔpK at 25°C for the acids oxalic to azelaic are compared with theoretical values calculated from the above assumptions, with ϵ given by the value for water. The values should eventually reach the theoretical limit of 0·60 (log 4) as r increases. The agreement of experimental and calculated values is excellent when n is four or greater. The model grossly underestimates the effect of the charge on the acidity of the acid anion when n is small; the local dielectric constant will undoubtedly be less than that of the solvent in this situation, and a more sophisticated model has been examined.[5]

An obvious extension of this electrostatic model explains the effect of dipolar substituents in a molecule on the acidity constant. For instance, whilst the pK of acetic acid is 4·76, the pK decreases to 2·58 in fluoroacetic acid. The carbon–fluorine bond is strongly polar in the sense shown

$$HO_2C \cdot \overset{\delta^+}{CH_2} - \overset{\delta^-}{F}$$

TABLE 9.2

ΔpK for $HO_2C \cdot (CH_2)_n \cdot CO_2H$

n	0	1	2	3	4	5	6	7
ΔpK(exp)	2·99	2·85	1·43	1·05	1·00	0·94	0·89	0·86
ΔpK(calc)	1·38	1·19	1·08	1·00	0·99	0·94	0·89	0·86

and the ionization of the proton is thereby made easier in fluoroacetic acid. An interesting set of measurements has been reported[6] on the series $^-O_3S \cdot (CH_2)_n \cdot CO_2H$. When n is small, the acid-strengthening effect of the dipoles in the sulphonate ion, which in an extreme structure can be represented

$$- \overset{2+}{S} \equiv O_3^{3-},$$

outweighs the opposite effect of the net negative charge. When $n = 1$ the pK of the acid is 4·05 which is 0·71 less than that of acetic acid. Such dipolar effects on various equilibria and rates are systematized by physical organic chemists under the heading of inductive effects. Our knowledge of molecular configurations, charge distributions in dipoles, and local dielectric constants is inadequate to permit profitable quantitative calculations.

It is tempting but erroneous, naively to extend qualitative notions of the inductive effect of substituents in a molecule to predict that the more electron-attracting is the atom to which a hydrogen atom is attached the more readily will the molecule ionize. It is true that acidity increases in the series NH_3, H_2O, HF or in the series PH_3, H_2S, HCl but hydrogen sulphide is a much stronger acid than water and hydrogen chloride than hydrogen fluoride ($pK = 3.17$), which is the opposite of what might have been expected from electronegatives or bond polarities. It is easier to understand how this happens if one imagines the dissociation to be split into two stages, viz. HA → H + A → H⁺ + A⁻. The effect which we have been discussing in this section now appears as an effect on the electron affinity of A (in solution). But if the atom to which the hydrogen is attached, is itself changed there may be an even greater and opposite effect on the homolytic bond dissociation energy. Inspection of tables of covalent

TABLE 9.3
pK_a *values for inorganic oxyacids*

$X(OH)_n$		$XO(OH)_n$		$XO_2(OH)_n$		$XO_3(OH)_n$	
$B(OH)_3$	9.2	$CO(OH)_2$	3.9	$NO_2(OH)$	−1.4	$ClO_3(OH)$	strong
$Si(OH)_4$	10	$H·CO(OH)$	3.7	$SO_2(OH)_2$	strong	$MnO_3(OH)$	strong
$Ge(OH)_4$	9	$NO(OH)$	3.3	$SeO_2(OH)_2$	strong		
$As(OH)_3$	9	$PO(OH)_3$	2.1	$H·SO_2(OH)$	0.4		
$Te(OH)_6$	7.6	$H·PO(OH)_2$	2	$ClO_2(OH)$	strong		
ClOH	7	$H_2PO(OH)$	1	$BrO_2(OH)$	strong		
BrOH	8.6	$AsO(OH)_3$	2.3	$IO_2(OH)$	0.8		
IOH	10	$TeO(OH)_2$	2.6				
		$ClO(OH)$	2.0				
		$IO(OH)_5$	1.6				

bond energies shows that at any rate for homolytic dissociation in the vapour phase, the change from oxygen to sulphur or from fluorine to chlorine drastically weakens the bond.

9.3. pK_a values of oxyacids

An electrostatic model also gives a qualitative explanation of some striking regularities in the strengths of simple inorganic oxyacids which have been often noted and given a variety of interpretations.[7] The successive acidity constants of polybasic acids differ from one another by units of about 5 in pK, e.g. phosphoric acid with $pK_1 = 2.15$, $pK_2 = 7.20$, $pK_3 = 12.31$. Obviously the more negative charge on an anion, the harder it is to remove a proton. The pK_1 values[8] fall into fairly clearly defined groups depending on the number of oxygen atoms with unattached hydrogen, as will be seen from Table 9.3. Some of the values quoted are very uncertain because of the instability of the species involved. What is noteworthy is the degree to which independent structural information on the acids and their salts is borne out by the column in which an acid falls, as for example in the cases of telluric, phosphorous, hypophosphorous and periodic acids. The position of boric acid is perhaps accidental as the Raman spectrum of borate solutions suggests[9] that the ion is $B(OH)_4^-$, in which case the dissociation of boric acid would involve the addition of OH^- rather than the loss of H^+. The value quoted for carbonic acid is the true value after correcting the conventional value for the fact that only 0·3 per cent of dissolved carbon dioxide is believed to be in the hydrated form.[10] The value for sulphurous acid is obtained by combining the conventional pK with a value for $[H^+][HSO_3^-]/[SO_2]$ recently obtained by relaxation spectrometry[11] (8.4). The general similarity of the Raman spectrum of the sulphite ion in water with that of the $H-PO_3^{2-}$ ion and in particular the observation[12] of a line ascribed to the H—S vibration strongly supports the structure $H-SO_3^-$ for the ion. The pK value for iodic acid suggests that in water it may be partly $IO(OH)_3$ rather than $IO_2(OH)$.

An electrostatic interpretation of the grouping of the oxyacids is possible.[7] The groups differ in the number of oxygen atoms over which the charge on the anion can be spread. The electrostatic free energy of a sphere of charge e and radius r in a medium of dielectric constant ϵ is $e^2/\epsilon r$; if the charge is spread over x spheres of the same radius the free energy is $e^2/\epsilon r x$. This gives $K'' = K' \exp(e^2/\epsilon r \Delta x kT)$ and taking the macroscopic dielectric constant of the solvent and $r = 0.75$ Å we obtain for water at 25°C, $\Delta pK = 4.1$ for $\Delta x = 1$ which is of the right order of magnitude. Another factor of importance is probably the stabilization of the anion by bond resonance to a degree which depends on the number of identical oxygen atoms available. This is the mesomeric effect of the physical organic chemists, and is the usual explanation given of the reason for the much greater acidity of carboxylic acids than alcohols. Of course there can be no doubt that the acidity of a species such as nitromethane ($pK = 10$) is due to the contribution to the anion resonance hybrid of forms such as

$$CH_2 : \overset{+}{N}O^{2-},$$

in which the charge has been removed from the carbon atom.

REFERENCES

1. BELL, R. P., *The Proton in Chemistry*, Methuen, London, 1959, p. 71.
2. PITZER, K. S., *J. Amer. chem. Soc.*, 1937, **59**, 2365 ; EVERETT, D. H. and WYNNE-JONES, W. F. K., *Trans. Faraday Soc.*, 1939, **35**, 1384.
3. BJERRUM, N., *Z. phys. Chem.*, 1923, **106**, 219.
4. ROBINSON, R. A. and STOKES, R. H., *Electrolyte Solutions*, Butterworths, London, 1959, p. 520 ; GANE, R. and INGOLD, C. K., *J. chem. Soc.*, 1931, 2153 ; JONES, R. H. and STOCK, D. I., *J. chem. Soc.*, 1960, 102.
5. KIRKWOOD, J. G. and WESTHEIMER, F. H., *J. chem. Phys.*, 1938, **6**, 506, 513.
6. BELL, R. P. and WRIGHT, G. A., *Trans. Faraday Soc.*, 1961, **57**, 1377.
7. BELL, R. P., *The Proton in Chemistry*, Methuen, London, 1959, p. 92.
8. BJERRUM, J., SCHWARZENBACH, G. and SILLÉN, L. G., " Stability Constants ", Part II, *Chem. Soc. Special Publ.*, 1958, No. 7.
9. EDWARDS, J. O., MORRISON, G. C., ROSS, V. F. and SCHULZ, J. W., *J. Amer. chem. Soc.*, 1955, **77**, 266.
10. WISSBRUN, K. F., FRENCH, D. M. and PATTERSON, A., *J. phys. Chem.*, 1954, **58**, 693.
11. EIGEN, M., KUSTIN, K. and MAASS, G., *Z. phys. Chem. (Frankfurt)*, 1961, **30**. 130.
12. SIMON, A. and WALDMAN, K., *Z. anorg. Chem.*, 1955, **281**, 135.

CHAPTER 10

STABILITY CONSTANTS

10.1. Introduction

The effort that has been expended on the measurement of stability constants is well exemplified by the 426 references to papers on chlorocomplexes alone in the compilation by Bjerrum, Schwarzenbach and Sillén.[1] Just as with protolytic equilibria, sophisticated theoretical models are intractable, but the data do show some interesting patterns which can be related to simple theoretical models.

Little will be said about redox equilibria. They are qualitatively consistent with what is known from preparative work about the stability of the various valence states of the elements, which as far as the metals are concerned correlate with the ionization potentials of the metal atoms in the gaseous phase. The stabilization* of particular valence states of a metal by certain ligands, e.g. Co(III) relative to Co(II) by ammonia, is related to the relative stability constants of the complexes formed by the two ions with the same ligand.

A discussion of stability constants conveniently starts by an examination of how far a simple electrostatic model is capable of explaining the values of the constants. This model treats the solvent as a continuous dielectric medium. Then we examine our scanty knowledge of the state of hydration of ions in aqueous solution. We next discuss the step-wise nature of complex formation, i.e. the tendency for successive attachment of several ligands to the same cation. In this respect as in many others the proton is atypical; once it has attached itself to a single base site it is only by comparatively weak hydrogen bonds that it sometimes binds additional ligands. There follow three sections of discussion of the general pattern of stability constants. As typical of the major classes of ligands we consider the halide ions, oxygen in its various forms (OH^- and oxyanions), amines, the sulphide ion and the cyanide ion. In their complex-forming tendencies the cations

---

* The word is commonly misused when what is being discussed is reactivity with a particular reagent rather than thermodynamic stability.

divide into three classes;[2] those with a rare-gas electronic configuration (d^0 cations), those with an outer sub-shell of ten d electron (d^{10} cations) and the transition metal cations (d^n cations). In these sections the concern is with general trends rather than exact pK values; as emphasized in previous chapters, these often depend on more or less arbitrary assumptions made in analysing the results. The pK_d and pQ_d values used (K_d and Q_d in mole l^{-1}) are for the first stage of complex formation and refer to aqueous solutions at 25°C. The chapter concludes by a discussion of the extra stability of chelate complexes.

The interpretation of values of $\Delta H^\ominus$ and $\Delta S^\ominus$ is not yet on a sufficiently firm basis to justify detailed discussion, but attention is drawn to 9.1. A number of regularities among $\Delta S^\ominus$ values have been found, and they provide qualitative information about changes of solvation on complex formation. For example, for the formation of the complex FeOH^{2+} the values of the thermodynamic functions in $\Delta G^\ominus = \Delta H^\ominus - T\Delta S^\ominus$ are[3] $\Delta G^\ominus = -16 \cdot 1$ kcal mole^{-1}, $\Delta H^\ominus = -3 \cdot 0$ kcal mole^{-1}, $T\Delta S^\ominus = -13 \cdot 2$ kcal mole^{-1}. In view of the discussion in 9.1 it can be said that the stability of the complex is due more to a very favourable partition function quotient than to ΔE_0. The favourable partition function quotient is associated with the ionic desolvation which accompanies complex formation.

10.2. The electrostatic model for complex ion equilibria

If the bond in a complex ion is adequately described by a simple electrostatic model, that is by coulombic interaction between spherical non-polarizable ions with charges z_+e and z_-e in a medium of dielectric constant ϵ, then the dissociation constant K of the ion-pair* is given by[4]

$$K^{-1} = 4\pi N \int_a^d \exp(z_+ \mid z_- \mid e^2/\epsilon rkT) r^2 dr \qquad (10.2.1)$$

where the integral limits extend from the separation of the ions in contact a to some distance d (Fig. 10.1). The problem is to assign a value to d. A logical procedure is to equate d to the ion-size parameter in the conductance and/or activity coefficient equation for the free ions used to calculate K. Bjerrum originally suggested[4] that when calculating osmotic coefficients for a model electrolyte of given a it was best to set $d = z_+ \mid z_- \mid e^2/2\epsilon kT$. At this distance, which has

---

* If a value of r in cm is inserted in 10.2.1 and the following equations, the units of K are, of course, mole ml^{-1}.

become known as the Bjerrum critical distance, the distribution function of ions of opposite charge around a central ion goes through a minimum. However, in testing whether an experimentally determined K accords with the predictions of the electrostatic model, there is no reason to suppose that the experimental method used will have set the maximum separation of ions counted as associated at this minimum.

Equation 10.2.1 is conveniently written in the form

$$K^{-1} = 4\pi N s^3 \int_{s/d}^{s/a} e^x x^{-4} dx \qquad (10.2.2)$$

with $s = z_+ |z_-| e^2/\epsilon kT$ and $x = s/r$; values of the integral

$$\int_2^b e^x x^{-4} dx$$

have been tabulated.[5]

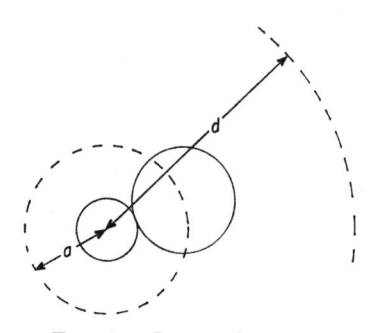

FIG. 10.1. Ion pair formation.

If in equation 10.2.1 $d - a \ll a$ the equation can be rewritten

$$K^{-1} = 4\pi N a^2 (d - a) \exp(z_+ |z_-| e^2/\epsilon akT) \qquad (10.2.3)$$

The arbitrary assumption that $d = 4a/3$ gives

$$K^{-1} = \frac{4}{3} \pi N a^3 \exp(z_+ |z_-| e^2/\epsilon akT) \qquad (10.2.4)$$

which is an equation favoured by Fuoss.[6] It is instructive to calculate from this approximate equation the value of K^{-1} which would result from the random collisions of uncharged molecules. If the exponential factor is unity and $a = 4$ Å, K^{-1} has a value of about 0·16 l mole^{-1} (a dissociation constant of about 6 mole l^{-1}).

It seems likely that the cases most favourable for the validity of the simple electrostatic model, with negligible interactions of a " chemical " nature, will be those where one or both partners are already stable complex ions. The conductimetric method gives[7] $pK = 3\cdot75_5$ for lanthanum ferricyanide, a value large enough to be rather insensitive to the conductance equation used or the ion-size parameter inserted therein (3.3). That the population of pairs of ions at intermediate distances of approach is in this case small is confirmed by the fact that when K is inserted into equation 10.2.2 and a evaluated after arbitrarily fixing d, the values are insensitive to d. They are $d = 32$ Å, $a = 7\cdot2$ Å; $d = 16$ Å, $a = 6\cdot9$ Å; $d = 10$ Å, $a = 6\cdot4$ Å. The values of a suggest that the associated species is $La(H_2O)_6^{3+}Fe(CN)_6^{3-}$. Unfortunately, direct crystallographic evidence about the effective radii of these ions is not available, but some reasonable and small corrections to known values for[8] $Co(NH_3)_6^{3+}$ and[9] $Mo(CN)_8^{4-}$ lead to $a = 3\cdot1$ Å $+ 4\cdot6$ Å $= 7\cdot7$ Å. The electrostatic model is not far removed from reality.

TABLE 10.1
Distances of closest approach from cryoscopic results

	$a/\text{Å}(d = 13\cdot9 \text{ Å})$	$a/\text{Å}(d = 6\cdot9 \text{ Å})$	$a/\text{Å}(d = 5 \text{ Å})$	$a/\text{Å}(d = 4 \text{ Å})$
$CuSO_4$	3·6	3·4	3·4	3·3
$ZnSO_4$	3·9	3·7	3·7	3·5
$MgSO_4$	3·9	3·7	3·6	3·5
$CaSO_4$	3·6	3·5	3·4	3·3
$CoSO_4$	4·0	3·7	3·7	3·5
$NiSO_4$	3·9	3·7	3·6	3·4

For the bi-bivalent sulphates discussed in 2.4, 3.3 and 6.2, K is strongly dependent on the assumptions made in obtaining the degrees of dissociation α. However, provided that corresponding values of K and d are inserted in 10.2.2 the values of a obtained therefrom are remarkably consistent, as shown in Table 10.1, and similar for the different sulphates. Their dependence on the exact value assumed for ϵ is small.[10] The a values are close to the average sulphur–metal distance of 3·64 Å in anhydrous sulphates.[11] The almost perfect agreement is, however, probably fortuitous for both the optical behaviour (2.4) and relaxation spectrometry (8.4) suggest that the complexes are predominantly outer-sphere. Furthermore, known inner-sphere complexes of the cations (e.g. oxalates) show large differences in stability. We shall return to this shortly.

Another example where the simple model seems to give approximately the right answer is that of $CaOH^+$. Several methods have given values of pK varying from 1·40 to 1·15 depending on the exact assumptions made in obtaining it.[12] If $K = 0.05$ mole l^{-1} is inserted in equation 10.2.1 with $d = 7.13$ Å ($s/2$) one obtains $a = 2.6$ Å and with $d = 3.57$ Å, $a = 2.0$ Å. The sum of the crystallographic radii is 0.99 Å $+ 1.4$ Å $= 2.4$ Å. However, extension to $MgOH^+$, for which p$K = 2.58$, gives $a = 1.3$ Å which is considerably smaller than 2.05 Å, the sum of the crystallographic radii.

It is instructive to consider now some of the possible causes of deviation from the predictions of the simple model. It neglects covalent interaction; to fit the pK of acetic acid the absurd value of $a = 0.3$ Å is required. It neglects the charge distribution in polyatomic ions. The interaction of a potassium ion with three fluoride atoms carrying a high negative charge density at one face of the octahedral PF_6^- ion may be the reason why there is a small but unambiguous degree of association of $K^+PF_6^-$ in aqueous solution with[13] $K = 0.41$ mole l^{-1}; the sum of the crystallographic radii is about 3.9 Å which is 0.4 Å larger than $s/2$. The electrostatic model neglects van der Waals attraction between ions. Data cited later (10.5) show that large anions display a surprising tendency to associate with large cations. The effects of specific free ion–solvent or complex ion–solvent interaction are also ignored. An ion will at least partly desolvate when complex formation occurs, but it is possible that in an outer-sphere complex a strongly polarized water molecule retained in the inner co-ordination sphere of the cation might form a particularly strong hydrogen bond with the anion thereby causing an extra electrostatic interaction not allowed for in the simple theory. This kind of interaction has been called[14] " localized hydrolysis ". It may explain why although the bi-bivalent sulphate complexes are outer-sphere the a values calculated from equation 10.2.1 correspond to inner-sphere distances. The six water molecules in the octahedral co-ordination sphere of the cation will be strongly polarized. Three of the fractionally positively charged hydrogen atoms at one face of the octahedron will be geometrically well situated to form strong hydrogen bonds with three of the oxygen atoms of a tetrahedral sulphate ion located in contact with the octahedron face. Various other sources of deviations from the simple electrostatic model will be discussed later. Calculations[10] show that the difficulties arising from ignorance of the exact value of the dielectric constant ϵ in equation 10.2.1 have often been exaggerated.

10.3. Ionic hydration

The immediate molecular environment of ions in a solvent is a topic on which much has been written and little is known with certainty. It was noted in 2.1 that spectroscopic evidence suggests that transition metal cations in water have six octahedrally attached water molecules. Isotopic labelling has been used[15] to show that the Cr^{3+} ion in aqueous solution has six water molecules attached to it which only exchange very slowly with the solvent. Fast reaction techniques can be used[16] to extend this kinetic criterion to other cases and the results are consistent with reasonable hydration numbers such as 6 for Al^{3+} and 4 for Be^{2+}. Spectrophotometric measurements[17] on the equilibrium $B + H^+(H_2O)_h \rightleftarrows BH^+ + hH_2O$ in solutions of strong acids are, with some rather drastic assumptions (no solvation of anions or BH^+ and equality of activity coefficients of univalent ions), consistent with $h = 4$. It is reasonable that the three hydrogen atoms of H_3O^+ are ready to form strong hydrogen bonds with other water molecules. Such diverse properties as the transport of water by ions when they diffuse, the size of the hydrodynamic unit inferred by applying Stokes' Law or some refinement of it to ionic mobilities, activity coefficients of concentrated solutions, and compressibilities, have all been used to estimate the number of water molecules in the ion–solvent complex.[18] The values are reasonable, but not very concordant, and there is no reason why they should be; the apparent hydration number of an ion depends on the experimental criterion used.

It is important to remember that complex formation relates to competition for the cation between the ligand and solvent molecules. The same competition for the cation between the solvent and a ligand also governs solubility equilibria and there are often parallelisms between such equilibria and complex ion equilibria in homogeneous solution. For instance, cations which form stable complexes with polarizable ligands also have very insoluble sulphides. The essential difference between precipitation and complex formation is that in a precipitate, in contrast with a mononuclear complex, each anion shares its attentions among several cations. This is true to a limited extent in polynuclear complexes in solution. Such polynuclear complexes are extensively formed as metastable intermediates in the hydrolysis of metallic cations, and their predominance makes the measurement of the stability of mononuclear hydroxy-complexes often difficult and sometimes impossible.[19]

10.4. Step stability constants

Before making a survey of stability constants it is convenient briefly to consider the relationship of consecutive step stability constants. Some typical examples[1] are given in Table 10.2. Termination of complex formation, or large decreases in stability constants, occur at co-ordination numbers characteristic of particular cations. These co-ordination numbers are familiar from preparative chemistry. The co-ordination number is unlikely to exceed six on steric grounds, but when it is two or four this can be explained,[20] as well as the stereochemistry of the complexes, in terms of the orbitals used for binding and of ligand field theory. Up to the characteristic co-ordination

TABLE 10.2
Successive pK_d and pQ_d values

Cation	Ligand	pK_1	pK_2	pK_3	pK_4	pK_5	pK_6
[a]Al^{3+}	F^-	6·13	5·02	3·85	2·74	1·63	0·47
[b]Ag^+	NH_3	3·23	3·83				
[c]Hg^{2+}	NH_3	8·8	8·7	1·0	0·8		
[b]Cu^{2+}	NH_3	3·99	3·34	2·73	1·97		
[b]Zn^{2+}	NH_3	2·18	2·25	2·31	1·96		
[b]Co^{2+}	NH_3	1·99	1·51	0·93	0·64	0·06	−0·74

[a] pQ_d; 0·53M KNO_3; 25°C
[b] 30°C
[c] pQ_d; 2M NH_4NO_3; 22°C

number the consecutive step stability constants usually decrease steadily. Even with a neutral ligand, and in the absence of steric congestion due to a bulky ligand, this will tend to happen because the number of sites at which replacement of a water molecule by a ligand can occur decreases as substitution proceeds. However, with a simple numerical allowance for this so-called statistical effect (cf. successive pK's of polybasic acids), there is usually a residual effect which must be electronic in origin. Nevertheless there are some cases where a striking reversal of the normal behaviour occurs. One example is the pK's for the silver ammines (Table 10.2). The comparative instability of the intermediate complex is probably due to the characteristic co-ordination number being different for water and for the substituting ligand which means that in the intermediate complex one of the ligands is necessarily in a complex with an uncongenial co-ordination number. Similarly a change of co-ordination number of the zinc ion from six for water to four for ammonia, which is consistent with the composition of solid co-ordination compounds, explains why the stability of the ammines increases with the addition of the first three ammonia molecules (Table 10.2). Abnormal instability of intermediate complexes can also be caused by variation of the electronic state of the central ion with the ligand. The hydrated ferrous ion is spin-free with four unpaired electrons but the complexes with three molecules of 2,2'-dipyridyl or 1,10-phenanthroline (both bidentate ligands) are spin-paired and diamagnetic. The intermediate complexes are relatively unstable.[1]

TABLE 10.3

pK_d and pQ_d values

	Li$^+$	Na$^+$	K$^+$	Mg^{2+}	Ca^{2+}	Sr^{2+}	Ba^{2+}
F$^-$				1·82	<1·0		<0·5
OH$^-$	−0·1			2·6	1·3	1·0	0·6
P$_3$O$_{10}^{5-}$	3·9	2·7	2·7	8·6	8·1	7·2	6·3
OAc$^-$				0·78	0·77	0·44	0·41
Ox^{2-}				3·4†	3·0†	2·5†	2·3
edta	2·7$_9$*	1·66*		8·69*	10·70*	8·63*	7·76*
NO$_3^-$		−0·6	−0·2		0·28	0·82†	0·92†
IO$_3^-$		−0·5	−0·3	0·82	0·89†		1·1
SO$_4^{2-}$	0·64†	0·70	0·82	2·2†	2·3†		
S$_2$O$_3^{2-}$		0·68	0·92	1·84	2·02	2·04	2·3

* pQ_d in 0·1M KCl at 20°C
† 18°C
edta denotes ethylenediaminetetra-acetate

In the subsequent analysis of the behaviour of the various classes of cations, values[1] of the first stability constant only will be discussed.

10.5. d^0 cations

A selection of stability constants for alkali metal and alkaline earth cations is given[1,21] in Table 10.3 (many of the values are sensitive to the method used and the assumptions made in analysing the results, but they are adequate for comparative purposes). The more highly charged ions form the stabler complexes, but the stability constants show two kinds of behaviour in their dependence on ion size. For ligands with concentrated negative charge, such as the fluoride ion and the anions of weak oxyacids, the stability constants increase as the cation becomes smaller, which agrees with the qualitative predictions of the simple electrostatic model (anomalies with chelate complexes, such as that of magnesium with ethylenediaminetetra-acetate, may be due to steric congestion). We also recall the strong tendency of Be^{2+} and Al^{3+} to form complexes with this type of ligand.

There is, however, an entirely different type of behaviour also exemplified by the data in Table 10.3. With oxyanions with a well-distributed charge, the stability of the complexes is low but shows an unambiguous increase with size of cation. It is noteworthy that the solubilities of alkali metal perchlorates and of alkaline earth sulphates follow this same sequence. Factors outside the scope of the simple electrostatic model discussed in 10.2 must be important here, and it is necessary to invoke the molecular nature of the solvent. As a model, consider a solvent molecule with a dipole of one unit of electronic charge 1 Å from a unit of protonic charge centred in a molecule of 2 Å diameter. This molecule would have a dipole moment of 4·8 debye, which is a not unreasonable value for a water molecule, with a permanent dipole moment of 1·8 debye, polarized in the intense field of an ion. The mutual potential energy of the dipole in contact with and fully oriented by a cation of radius r_c can be evaluated by elementary electrostatics and is given by

$$U = \left(\frac{1}{r_c + 1\cdot 5\,\text{Å}} - \frac{1}{r_c + 0\cdot 5\,\text{Å}}\right)\frac{e^2}{\epsilon} = \frac{\phi e^2}{\epsilon} \qquad (10.2.5)$$

where ϕ is plotted in Fig. 10.2. The corresponding values of ϕ for electrostatic interaction of the cation with iodide and fluoride ions are also given. It is clear from the figure that in the absence of specific factors large anions will be unable to displace water from the inner co-ordination sphere of small cations. Exactly the same considerations will apply to the ease of desolvation of anions. At this point it is interesting to note that equations 10.2.1 or 10.2.4 predict a dissociation constant of about unity for sodium chloride if a is given by the sum of the crystallographic radii. A dissociation constant of less than ten is also predicted[22] by an elementary statistical mechanical calculation starting with equation 9.1.1. Only by invoking the molecular nature of the solvent is the complete dissociation of even alkali metal and alkaline earth halides explicable. On the other hand, the statistical mechanical calculation[22] predicts a much readier dissociation of electrolytes with polyatomic ions, because of a more favourable partition function quotient. It is ease of desolvation reinforced by van der Waals attraction that is probably responsible for the association of pairs of large ions typified by the data in Table 10.3, caesium iodide (6.3) and tetra-alkylammonium salts with large anions.[23]

The results of competition for the company of the cation between the anion and water shows clearly in the activity coefficients[24] of salts of d^0 cations. With large anions such as Cl^-, Br^-, I^-, NO_3^-, ClO_3^-, ClO_4^- activity coefficients diminish with increasing size of cation for both alkali metal and alkaline earth salts. The larger the cation the more

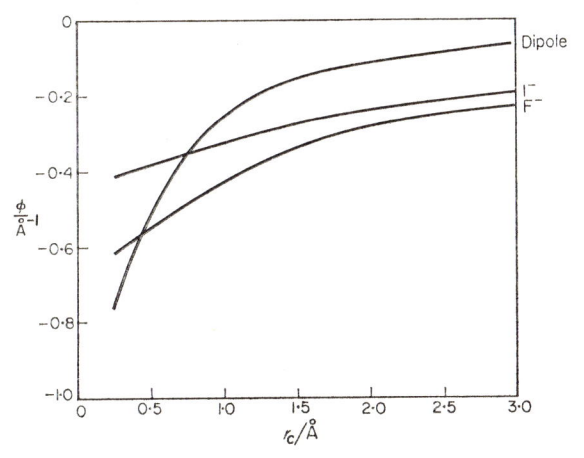

Fig. 10.2. Mutual potential energy of interaction of F^-, I^- and a model dipole with a cation of radius r_c.

readily it associates with anions with at the same time a reduction in the solvation of both ions. The effect of both factors is to decrease the activity coefficient of the salt. With salts of the hydroxyl and carboxylate ions the order of activity coefficients is reversed; here the smaller the cation the more it associates with the anion and the lower the activity coefficient. Salts of d^0 cations therefore have abnormally low activity coefficients when *both* ions are either large or small (6.3). Of course in concentrated solution, association will not be restricted to the binary stage alone, but the qualitative effects will remain the same.

The same pattern is clearly discernable in the solubilities of alkali metal salts, which fall almost entirely into two distinct classes.[25] With large anions of strong acids the order of solubilities is Li > Na > K > Rb > Cs, whilst with the anions of weak acids the sequence is reversed. With the alkaline earths the solubility pattern[25] is less clear cut. The sequence is usually Ca > Sr > Ba, but the reverse order is found for the hydroxides, fluorides and oxalates. The deviation of the solubilities of other carboxylates from the expected sequence and the rise of solubility of fluorides and oxalates in the order Ca < Mg < Be is probably connected with factors such as the stability of soluble complexes and steric congestion in the solid lattices.[26]

In the interpretation of activity coefficients it has sometimes[14] been held that the complexes of d^0 cations with the anions of weak acids are of the outer-sphere type. Polarization of water molecules retained by a cation will increase with z_+/r_c and the localized hydrolysis

TABLE 10.4
pQ_d values in 3M NaClO$_4$

	F$^-$	Cl$^-$	Br$^-$	I$^-$
Zn^{2+}	0·73*	−0·19	—	—
Cd^{2+}	0·57	1·4	1·8	2·1
Hg^{2+}	1·0*	6·7*	9·0*	12·9*

* 0·5M NaClO$_4$

interaction (10.2) with an associated anion will increase in parallel. It is difficult to discriminate between the two alternative explanations, and both types of complex may be present in some cases, but the way in which the parallelism can be extended to include anhydrous precipitates is evidence against the localised hydrolysis explanation.

10.6. d^{10} cations

In complete contrast to the d^0 cations, formation of stable soluble complexes or insoluble precipitates occurs with chloride, bromide, iodide, sulphide and cyanide ions and with ammonia. It is ions with this electronic configuration (or with an additional "inert" pair of electrons giving $d^{10} + s^2$) that have sulphides insoluble in acid solution, except that Cu^{2+} replaces Zn^{2+} in this context. Zinc also shows only slight tendency to form complexes with halide ions. A comparison of pQ_d values for the first complexes of zinc, cadmium and mercury with halide ions is interesting (Table 10.4). The sequence of stabilities of the zinc complexes with F$^-$ and Cl$^-$ is the one that the electrostatic model would predict, but the opposite is the case for cadmium and mercury. One expects covalent interaction to become progressively more important as the polarizability of the anion and the polarizing power of the cation increase. The former will certainly increase in the order $F^- < Cl^- < Br^- < I^-$. Some approximate values[1,27] for the stability constants of silver complexes* in Table 10.5 show a rough parallelism with qualitative notions of anion polarizability. However, the iodide complex with the singly charged Ag^+ ion is six pK units more stable than the iodide complex with the doubly-charged Cd^{2+} ion. Cyanide, thiocyanate and thiosulphate complexes show the same trend. Some factor quite other than the polarizability of the anion or the polarizing power of the cation must also be important. This is possibly bonding in which filled d orbitals on the cations interact with available vacant orbitals on the ligand atoms. Such binding is well established in

TABLE 10.5

pK_d and pQ_d values for silver complexes

NO_3^-	OAc^-	OH^-	SO_4^{2-}	SH^-	S^{2-}	NH_3	F^-	Cl^-	Br^-	I^-
−0·2	0·7	2·0	1·3	13·6*	20·3*	3·3	−0·4	3·3	4·4	8·1†

* 0·1M $NaClO_4$
† 4M $NaClO_4$

carbonyl, cyanide and other low-valency complexes of transition elements. This kind of binding cannot happen with ammonia as a ligand and the much smaller difference in the same sense between a pK_d of 2·5 for $Cd(NH_3)^{2+}$ and of 3·3 for $Ag(NH_3)^+$ is probably due to the effect of a relatively firmer attachment of water by Cd^{2+}. Kinetic studies by relaxation techniques show[28] that the rate constants for complex formation by d^{10} cations, which are governed by the removal of a water molecule, are similar to those for alkaline earth cations. This strongly suggests that differences in the interactions with water molecules are not an important factor in determining the differences between stabilities of complexes of d^0 and d^{10} cations.

Divalent lead and univalent thallium only differ electronically from Hg^{2+} in the possession of an extra pair of electrons (the " inert " pair). The complexes of these cations have been rather thoroughly investigated. They form much weaker complexes than Hg^{2+}, but the general sequence of stabilities is similar (Table 10.6), although their affinity

---

* Silver halides are often thought of as strong electrolytes; in many contexts the constant concentration of AgX in saturated solutions can be ignored.

for amines is slight. The thallous ion is remarkable as a univalent ion with pronounced complex-forming tendencies. The contrast with rubidium, of almost identical radius is striking; again the electrostatic model fails. It is likely that in an unsymmetrical electric field the 6s electron pair moves to a 6sp orbital, leaving a vacant 6sp orbital on the opposite side of the cation for anion co-ordination.[29]

10.7. d^n cations

Quantitative measurements are largely confined to complexes of members of the first transition series; in the later series reactions are slow which makes equilibrium measurements difficult. Thermodynamic and conductance data show that the interaction of chloride, bromide

TABLE 10.6

pK_d and pQ_d values

	F$^-$	Cl$^-$	Br$^-$	I$^-$	NO$_3^-$	SO$_4^{2-}$	OH$^-$	NH$_3$
Hg^{2+}	1·56	6·74*	8·95*	12·87*	0·11†	1·34*	10·3*	8·8
Pb^{2+}	<0·8	1·57	1·77	1·92	1·15	—	6·2	—
Tl$^+$	−0·1	0·51	1·0	0·72‡	0·33	1·4	0·42	—

* 0·5M NaClO$_4$
† 3M NaClO$_4$
‡ 4M NaClO$_4$

or iodide ions with divalent cations from manganese to copper is weak (the chemical reaction of cupric and iodide ions is an exception). The occurrence of very intense absorption bands such as those associated with CuBr$^+$ sometimes gives an impression that the contrary is the case. The fluoride ion and the oxyanions of strong acids interact to about the same extent as with alkaline-earth cations. The oxyanions of weak acids, cyanide and sulphide ions and amines show strong interaction. In addition to the factors so far discussed there is another one of importance in its influence on the stability of transition metal complexes. This is crystal field stabilization.[20] For example, the field of six ligands disposed octahedrally around a cation has the effect of splitting the energetically degenerate five 3d orbitals in the isolated ion into a doublet and triplet, the doublet being at a higher energy level. In this group the electron density is greatest close to the ligands, whilst in the triplet group it is directed between them. If the d-subshell is incomplete the system can achieve a lower energy state by filling the d-orbitals other than randomly. The extra stabilization thereby achieved is known as crystal field stabilization and can be quantitatively assessed from spectrophotometric measurements of the spacing of the split d levels. Some ligands, for example ammonia and cyanide ion, exert a more powerful ligand field than water molecules and therefore when exchange of ligands occurs on complex formation extra stabilisation is achieved.* In an octahedral complex of the spin-free type (i.e. ligand field splitting not so large that more electrons are paired than in the free ion) the stabilization energy is zero at $d^5(Mn^{2+})$

TABLE 10.7

pK_d and pQ_d values

		Mn^{2+}	Fe^{2+}	Co^{2+}	Ni^{2+}	Cu^{2+}	Zn^{2+}
*NH_3	pK_1			1·99	2·67	3·99	2·18
	pK_2			1·51	2·12	3·34	2·25
	pK_3			0·93	1·61	2·73	2·31
	pK_4			0·64	1·07	1·97	1·96
	pK_5			0·06	0·63		
	pK_6			−0·74	−0·09		
†$(NH_2 \cdot CH_2)_2$	pK_1	2·73	4·28	5·89	7·52	10·55	5·71
	pK_2	2·06	3·25	4·83	6·28	9·05	4·66
	pK_3	0·88	1·99	3·10	4·26		1·72
$CH_2(CO_2^-)_2$	pK_1	3·29		3·72	4·1	5·60	3·68

* 30°C
† 1M KCl (Cu^{2+} : 0·5M KNO_3) at 30°C

and $d^{10}(Zn^{2+})$ and rises to a maximum at $d^8(Ni^{2+})$. The polarizing power of the cations, in so far as one can judge from the sum of the first two ionization potentials of the gaseous atoms, increases steadily from manganese to copper and then falls at zinc, and the ionic radii decrease slightly from Mn^{2+} to Cu^{2+} and increase again with Zn^{2+}. It is not surprising therefore that for (inner-sphere) complexes of this series the sequence of stability constants (often known as the Irving–Williams[30] order) is almost invariably $Mn^{2+} < Fe^{2+} < Co^{2+} < Ni^{2+} < Cu^{2+} > Zn^{2+}$. Some illustrative data[1] are collected in Table 10.7. It is also noteworthy that the insolubility of (aged) sulphides and hydroxides follows the same sequence. Most of the known exceptions to this rule are understood. Cations with a co-ordination number of four are

* This is, of course, an effect on ΔE_0 of equation 9.1.1.

naturally at a disadvantage when five or six ligands are attached. The preference of copper for four firmly held groups in a plane can give rise to stereochemical difficulty with polydentate ligands (see 10.8). The surprising stability of divalent iron complexes with some heterocyclic bases is thought to be due to the interaction of d electrons of the metal with vacant π orbitals of the ligand.

The ubiquitous tendency to complex formation of the more highly charged transition metal cations is well known. Indeed the Co(III) state is only known in aqueous solution when stabilized by complex formation with a ligand other than water. Careful studies have been reported of the stability of the complexes of Fe(III) with ClO_4^- and Cl^-. In the perchlorate complex[31] the ions are separated by two water molecules if the simple electrostatic model is correct, whilst relaxation spectrometry[32] (8.4) shows that the first chloro-complex is equally divided between inner- and outer-sphere states. Comparison of the value[31] $pK_d = 11\cdot82$ for $FeOH^{++}$ with that of $1\cdot85$ for $Co(NH_3)_6^{++}$ OH^- shows that it is an inner-sphere complex, and also how overriding non-electrostatic factors become with the more strongly complex-forming ligands.

10.8. Chelate stability

A matter that has not been explicitly discussed so far is the reason for the high stability of chelate complexes in which a ligand has two or more functional groups capable of attachment to a cation. The contrast in the pK_1 values for ammonia and ethylenediamine in Table 10.7 is notable. In the second case, provided the molecule formed is strain free, the ligand can attach itself by two bonds to the cation and yet the translational freedom of only one molecule has been lost, just as in the addition of a single ammonia molecule. This simple argument implies that ΔE_0 in equation 9.1.1 is changed favourably and the partition function quotient unchanged. With polyamines, the free energy of formation of a complex contains an approximately constant contribution for each functional group in the ligand. For example, the average contribution to pK per amino group for the attachment of two ethylenediamine molecules to the cupric ion is about 5 (Table 10.7). The pK_d for the complex with $(NH_2\cdot CH_2\cdot CH_2)_2NH$ is[1] $16\cdot0$ and with $(H_2N\cdot CH_2\cdot CH_2\cdot NHCH_2)_2$ is $20\cdot4$. The complex with $(NH_2\cdot CH_2\cdot CH_2)_3N$ which is less well suited to a square planar configuration around the copper ion is less stable ($pK_d = 18\cdot8$). Our crude argument has neglected degrees of freedom other than those of translation. The restrictive effect of chelation on the flexibility of a long molecule is no doubt the reason why chelate stability decreases as the ring size increases above five members, whilst the decrease in the effect for rings less than four-membered is due to steric strain.

The most powerful general complexing-agent known is the ethylenediaminetetra-acetate ion, $(^-O_2C \cdot CH_2)_2 \cdot N \cdot CH_2 \cdot CH_2 \cdot N(CH_2 \cdot CO_2^-)_2$. The geometry of the molecule is such that it can, although there is evidence[33] that it often does not, occupy all six positions in the co-ordination sphere of a cation with the formation of five 5-membered chelate rings. It has a high negative charge and both amino and carboxylate groups. It thus caters for all types of cations. Some pQ_d values[1] are given in Table 10.8. It even forms complexes with alkali metal cations. The higher stability of the calcium complex compared with that of

TABLE 10.8
pQ_d values for ethylenediaminetetra-acetate ion complexes

1.	H^+ 10·26	Li^+ 2·79	Na^+ 1·66	Mg^{2+} 8·69	Ca^{2+} 10·70	Ba^{2+} 7·76	La^{3+} 15·13
2.	Ag^+ 7·32	Zn^{2+} 16·50	Cd^{2+} 16·46	Hg^{2+} 21·80	Pb^{2+} 18·04		
3.	Mn^{2+} 13·58	Fe^{2+} 14·33	Co^{2+} 16·21	Ni^{2+} 18·56	Cu^{2+} 18·79	Fe^{3+} 25·1	

1 and 3 at 20°C in 0·1M KCl
2 at 20°C in 0·1M KNO_3

magnesium is presumably due to steric congestion around the smaller cation and is usual[26] with complexes of magnesium with polydentate ligands. The expected sequence of stabilities occurs for the divalent transition-metal cations.

REFERENCES

1. BJERRUM, J., SCHWARZENBACH, G. and SILLÉN, L. G., " Stability Constants ", Parts I and II, *Chem. Soc. Special Publ.*, 1957, No. 6 ; 1958, No. 7.
2. SCHWARZENBACH, G., *Recent Advances in Inorganic Chemistry and Radiochemistry*, Ed. EMELEUS, H. J. and SHARPE, A. G., Vol. 3, Academic Press, New York, 1961, p. 257.
3. MILBURN, R. M., *J. Amer. chem. Soc.*, 1957, 79, 537.
4. BJERRUM, N., *Kgl. danske Videnskab. Selskab, Mat.-fys. Medd.*, 1926, 7, No. 9.
5. GUGGENHEIM, E. A., *Faraday Soc. Discussions*, 1957, 24, 53 ; ROBINSON, R. A. and STOKES, R. H., *Electrolyte Solutions*, Butterworths, London, 1959, p. 549.
6. FUOSS, R. M., *J Amer. chem. Soc.*, 1958, 80, 5059.

7. GUGGENHEIM, E. A. and PRUE, J. E., *Physicochemical Calculations*, North-Holland, Amsterdam, 1956, p. 366.
8. CATON, J. A. and PRUE, J. E., *J. chem. Soc.*, 1956, 671.
9. HOARD, J. L. and NORDSIECK, H. H., *J. Amer. chem. Soc.*, 1939, **61**, 1853.
10. ROSSEINSKY, D., *J. chem. Soc.*, 1962, 785.
11. KOKKOROS, P. A. and RENTZEPERIS, P. J., *Acta cryst.*, 1958, **11**, 361.
12. BATES, R. G., BOWER, V. E., CANHAM, R. G. and PRUE, J. E., *Trans. Faraday Soc.*, 1959, **55**, 2062.
13. ROBINSON, R. A., STOKES, J. M. and STOKES, R. H., *J. phys. Chem.*, 1961, **65**, 542.
14. ROBINSON, R. A. and HARNED, H. S., *Chem. Rev.*, 1941, **28**, 419.
15. HUNT, J. P. and TAUBE, H., *J. chem. Phys.*, 1950, **18**, 757 ; 1951, **19**, 602.
16. JACKSON, J. A., LEMONS, J. F. and TAUBE, H., *J. chem. Phys.*, 1960, **32**, 553.
17. BELL, R. P. and BASCOMBE, K. N., *Disc. Faraday Soc.*, 1957, **34**, 158.
18. ROBINSON, R. A. and STOKES, R. H., *Electrolyte Solutions*, Butterworths, London, 1959, p. 52.
19. SILLÉN, L. G., *Quart. Rev. chem. Soc., Lond.*, 1959, **13**, 146.
20. ORGEL, L. E., *An Introduction to Transition-Metal Chemistry : Ligand Field Theory*, Methuen, London, 1960, p. 53.
21. DAVIES, C. W., *The Structure of Electrolytic Solutions*, Ed. HAMER, W. J., John Wiley, New York, 1959, p. 19 ; WOLHOFF, J. A. and OVERBEEK, J. T. G., *Rec. Trav. chim. Pays-bas*, 1959, **78**, 759.
22. FOWLER, R. H. and GUGGENHEIM, E. A., *Statistical Thermodynamics*, Cambridge, 1949, p. 375.
23. BOWER, V. E. and ROBINSON, R. A., *Trans. Faraday Soc.*, 1963, **59**, 1717.
24. ROBINSON, R. A. and STOKES, R. H., *Electrolyte Solutions*, Butterworths, 2nd ed., London, 1959, pp. 491–498.
25. SIDGWICK, N. V., *The Chemical Elements and their Compounds*, Vol. 1, Oxford, 1950, pp. 96, 219.
26. WILLIAMS, R. J. P., *J. chem. Soc.*, 1952, 3770.
27. ANTIKAINEN, P. J., HIETANEN, S. and SILLÉN, L. G., *Acta chem. Scand.*, 1960, **14**, 95 (AgOH) ; ZÜST, H., GÜBELI, O. and SCHWARZENBACH, G., *Chimia*, 1958, **12**, 84 (AgS$^-$ and AgSH).
28. EIGEN, M., *Pure and applied Chem.*, 1963, **6**, 97.
29. PANCKHURST, M. H., *Austral. J. Chem.*, 1962, **15**, 194.
30. IRVING, H. and WILLIAMS, R. J. P., *Nature, Lond.*, 1948, **162**, 746.
31. RICHARDS, D. H. and SYKES, K. W., *J. chem. Soc.*, 1960, 3626.
32. WENDT, H. and STREHLOW, H., *Z. Elektrochem.*, 1962, **66**, 226.
33. HIGGINSON, W. C. E., *J. chem. Soc.*, 1962, 2761.

CHAPTER 11

NON-AQUEOUS AND MIXED SOLVENTS

11.1. Simple electrostatic model

The simplest conceivable state of affairs is one in which the variation of an equilibrium constant with solvent is a function solely of solvent dielectric constant. For an ion-pair, $z_+ \mid z_- \mid e^2/\epsilon$ determines the extent of association, and a uni-univalent salt in a solvent with a dielectric constant of about 20 (e.g. acetone or propanol) will be associated to about the same degree as a bi-bivalent salt in water. To test the simple electrostatic model, studies of the association of alkali halides are suitable. Dissociation constants from conductance measurements have been reported for potassium iodide dissolved in pyridine[1], sulphur dioxide,[2] ammonia,[3] acetone,[4] methyl ethyl ketone,[5] n-propanol,[6] ethanol[7] and ethanolamine.[8] This covers a dielectric constant range of 12 to 38. Potassium chloride has been studied[9] in dioxan + water mixtures over the same dielectric constant range. It is likely that specific " chemical " interactions of the solvent with the large potassium and halide ions will be small. The electrostatic model predicts that the dissociation constant K will be given by equation 10.2.1 which for a uni-univalent salt is

$$K^{-1} = 4\pi N \int_a^d \exp(e^2/\epsilon r k T) r^2 \mathrm{d}r \qquad (11.1.1)$$

A simplified version (10.2) is

$$K^{-1} = \frac{4}{3}\pi a^3 N \exp(e^2/a\epsilon k T) \qquad (11.1.2)$$

where $4\pi a^3/3$ is the volume around a cation available to the centre of a paired anion. This equation predicts that pK will be a linear function of $1/\epsilon T$, and a plot of the two quantities is shown in Fig. 11.1. The results are for a temperature of 25°C except those for ammonia (−33·3°C) and sulphur dioxide (0·12°C). The pK values for potassium iodide were calculated using either the limiting law or the Shedlovsky equation (3.2.7) for the conductance of the free ions. The pK value in n-propanol decreases from 2·52 to 2·36 if recalculated[10] using the simplified Fuoss–Onsager equation (3.2.13) with $d = 3\cdot 6$ Å. Only the result for ethanolamine is likely to be substantially more sensitive to the change, and as the results are of variable precision, and the correct form of conductance equation still uncertain (3.2), they have not been recalculated. Those for potassium chloride are based on application of the simplified Fuoss–Onsager equation. The pK values for both halides are approximately a smooth function of dielectric constant. As discussed earlier (10.2), it is not known how close a pair of oppositely

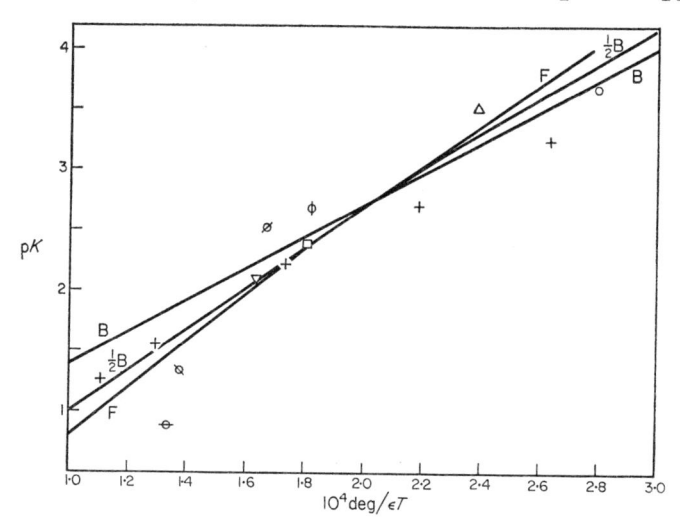

Fig. 11.1. pK of potassium iodide in pyridine (o), sulphur dioxide (△), ammonia (□), acetone (▽), methyl ethyl ketone (ϕ), n-propanol (ø), ethanol (⊗) and ethanolamine (⊖); pK of potassium chloride (+) in dioxan + water mixtures.

charged ions need to be before they cease to make a conductance contribution. One reasonable assumption is that this critical separation is the distance at which the mutual electrical potential energy of the pair of ions is equal to $2kT$, that is $d = e^2/2\epsilon kT$. This critical distance is that originally suggested by Bjerrum[11] for adoption in calculating the thermodynamic properties of a model electrolyte. With $a = 4\cdot 25$ Å, equation 11.1.1 gives the curve labelled B in Fig. 11.1. We note that with this assumption the value of d is a function of ϵ. This is not the case if, as Fuoss[12] prefers, equation 11.1.2 is used. This for $a = 4\cdot 25$ Å gives the curve labelled F in Fig. 11.1. Another reasonable assumption would be that at the critical distance the mutual potential energy of the pair of ions is equal to $4kT$, that is $d = e^2/4\epsilon kT$. This gives for $a = 4$ Å the line labelled $\tfrac{1}{2}$B in the figure. The curvature of this line seems to reflect the general trend of the results, but a certain answer to the precise location of d will need both experimental results of the highest precision and a theoretical conductance equation of assured accuracy. Then, the value of d used in the conductance equation should be consistent with that required in equation 11.1.1 to give the observed value of K. A good fit of the results for potassium chloride in dioxan + water is obtained with B($a = 5$ Å).

For these two halides therefore, the simple electrostatic model is approximately consistent with the results but there are discrepancies in the detail. The a values are all larger than the sums of the crystallographic radii which are 3·49 Å for KI and 3·14 Å for KCl. The curve for potassium chloride is below rather than above that for potassium iodide, and the individual points for potassium iodide in different solvents scatter markedly from a smooth curve. Specific solvation of one or both ions in the free and possibly in the associated states is a likely cause. Solvent sorting in the ionic fields is a related source of complication in the mixed solvent.

11.2. Specific solvation effects

The alkali halides were chosen in 11.1 in the belief that specific solvation effects would be small, as they seem to be. An extreme example of the contrary situation is afforded by the behaviour of aluminium bromide. Conductance measurements[13] show this to be a strong 3–1 electrolyte in dilute solutions in pyridine with a dielectric constant of 12, whilst it is a very weak electrolyte in nitrobenzene with a dielectric constant nearly three times as large. Solvation of the aluminium ion by the donor pyridine molecule must be responsible for this.

The effects of cation solvation on pK values are apparent even for alkali metal salts. From the results in Table 11.1 it is seen that for the picrates in the donor solvents pyridine and acetone the pK values are surprisingly small compared with those in nitrobenzene; furthermore, with the donor solvents the difference in the pK values for the three cations are much smaller than with nitrobenzene. This suggests that in these solvents the smaller the cation, the more firmly it is solvated. With the chlorides in ethanol, the interaction of the chloride ion with the cation is inadequate to counter the effects of firmer cation solvation as cation size diminishes, and the order of the dissociation constants is the reverse of that expected from the simple electrostatic model. This recalls a similar inversion in the association sequence of d^0 cations with large anions in water (10.5).

Pronounced solvation of the chloride ion in alcohols but not in dimethyl formamide is suggested by an increase in the pK of lithium chloride from[10] 1·4 to[15] 2·4 on changing the solvent from ethanol with a dielectric constant of 24·3 to dimethyl formamide which has a dielectric constant of 36·7. This is reasonable on structural grounds for in dimethyl formamide the positive end of the molecule is protected by two methyl groups; it is also supported[16] by a good deal of general evidence from solubilities and reactivities. More direct information about solvation and its firmness comes[15] from the calculation of hydrodynamic ion sizes from Λ_0 values and comparison of these with the values of the mean ionic diameter d necessary to fit conductance data for strong electrolytes by the Fuoss–Onsager equation. In this way it has been concluded that alkali metal cations but not halide ions

TABLE 11.1

pK values in non-aqueous solvents at 25°C

	Pyridine[1]	Acetone[4]	Nitrobenzene[14]		Ethanol[10]
ϵ	12·3	20·7	34·9	ϵ	24·3
LiPi	4·08	2·99	7·22	LiCl	1·43
NaPi	4·37	2·87	4·55	NaCl	1·64
KPi	4·00	2·46	3·16	KCl	1·98

are solvated in dimethyl formamide and in dimethyl sulphoxide, whilst in methanol both are solvated, but that penetration of the solvation sheath occurs on ionic encounter.

In protolytic equilibria, specific solvation effects usually exert a dominant influence. The basicity of solvent molecules is obviously important in governing acid ionization. For example, perchloric acid which is a very strong acid in the basic solvent water which has a dielectric constant of 78·5, has a pK of about 4 in sulphuric acid[17] which has a dielectric constant of 110. Acids are usually undissociated in hydrofluoric acid even though the dielectric constant is 84. Anion solvation is also important. Hydrobromic acid is very strong in water and picric acid has a pK of about 0·2; in dimethyl formamide the pK values are[18] 1·8 and 1·2 respectively, hydrobromic acid being the weaker acid in this solvent. If the dielectric constant is varied by changing the composition of a mixed solvent, or if the effects of proton solvation are eliminated by considering relative acidity constants, approximately linear relationships between pK or ΔpK and $1/\epsilon$ are found[19] for hydroxylic solvents, but the slopes of the plots are of no immediate theoretical significance.

11.3. Tetra-alkylammonium salts

There are many values reported in the literature of dissociation constants of tetra-alkylammonium salts in both pure and mixed non-aqueous solvents.[20] In solvents of dielectric constant less than about 30, the constants are not sensitive to the conductance equation used to obtain them and usually conform approximately to the predictions of the simple electrostatic model. For instance, the pK of tetrabutylammonium bromide in acetone is 2·48, the corresponding value of $a(d = s/2)$ being 3·5 Å which seems rather small, although the exact size and configuration of the tetrabutylammonium ion is uncertain. Changes of pK with size of cation or anion are as expected, but the specific influence of solvation is apparent with change of solvent. The pK of tetrabutylammonium picrate changes from 3·65 to 2·89 with a change of solvent from ethylene chloride ($\epsilon = 10\cdot23$) to pyridine ($\epsilon = 12\cdot01$); the corresponding values of $a(d = s/2)$ are 7·9 Å and 5·8 Å.

Measurements on solutions in benzene, which has a dielectric constant of 2·27 are of considerable interest. Association is measurable at concentrations so low that ion atmosphere effects are negligible. Below about 5×10^{-6} M only ion-pairs and free ions are present, but above this concentration triple and quadruple ions appear, the formation of which can also be treated in terms of simple ideal equilibria. Eventually higher aggregates appear, the formation of which has been studied cryoscopically; the degree of polymerization increases up to about 0·1M above which it decreases due to the increasing importance of long-range interionic forces. The pK of tetrabutylammonium bromide in benzene is 16·85 which corresponds to $a = 5\cdot9$ Å. From the dielectric constant of the solution the dipole moment of the ion-pair can be calculated; this when inserted in the equation $\mu = ea$ gives an independent value of 2·4 Å for a. Again the agreement is reasonable if not exact. The specific role hydrogen bonding can play in enhancing the stability of an associated species is shown by a dramatic increase of pK from 16·7 for tetra-amylammonium picrate to 20·4 for triamylammonium picrate.

REFERENCES

1. BURGESS, D. S. and KRAUS, C. A., *J. Amer. chem. Soc.*, 1948, **70**, 706.
2. LICHTIN, N. N. and LEFTIN, H. P., *J. phys. Chem.*, 1956, **60**, 161.
3. HNIZDA, V. F. and KRAUS, C. A., *J. Amer. chem. Soc.*, 1949, **71**, 1565.
4. REYNOLDS, M. B. and KRAUS, C. A., *J. Amer. chem. Soc.*, 1948, **70**, 1709.
5. HUGHES, S. R. C., *J. chem. Soc.*, 1956, 998.
6. GOVER, T. A. and SEARS, P. G., *J. phys. Chem.*, 1956, **60**, 330.
7. BRUSSET, H. and KIKINDAI, M., *Bull. Soc. chim. Fr.*, 1962, 1150.
8. BREWSTER, P. W., SCHMIDT, F. C. and SCHAAP, W. B., *J. Amer. chem. Soc.*, 1959, **81**, 5532.
9. LIND, J. E. and FUOSS, R. M., *J. phys. Chem.*, 1961, **65**, 999.
10. KAY, R. L., *J. Amer. chem. Soc.*, 1960, **82**, 2099.
11. BJERRUM, N., *Kgl. danske Videnskab. Selskab, Mat.-fys. Medd.*, 1926, **7**, No. 9.
12. FUOSS, R. M., *J. Amer. chem. Soc.*, 1958, **80**, 5059.
13. JACOBER, W. J. and KRAUS, C. A., *J. Amer. chem. Soc.*, 1949, **71**, 2405.
14. WITSCHONKE, C. A. and KRAUS, C. A., *J. Amer. chem. Soc.*, 1947, **69**, 2472.
15. PRUE, J. E. and SHERRINGTON, P. J., *Trans. Faraday Soc.*, 1961, **57**, 1796.
16. PARKER, A. J., *Quart. Rev. Chem. Soc., Lond.*, 1962, **16**, 163.
17. GILLESPIE, R. J., *J. chem. Soc.*, 1950, 2537.
18. SEARS, P. G., WOLFORD, R. K. and DAWSON, L. R., *J. Electrochem. Soc.*, 1956, **103**, 633.
19. BELL, R. P., *The Proton in Chemistry*, Methuen, London, 1959, p. 51.
20. KRAUS, C. A., *J. phys. Chem.*, 1956, **60**, 129.

INDEX

Absorbance 11
Absorption
  coefficient 11
  spectra of ions 9
Acetic acid 2, 25, 26, 39, 94
Acid–base
  equilibria 12
  indicators 12
Acidity constant 6, 82
Activity coefficients 4, 5, 36, 65, 99
Alkali halides in ethanol 34
Aluminium bromide 109
Ammonia 87
Ammonium nitrate 19
Ammonium sulphate 19
Amperometry 72
Association constant 6, 82

Barium hydroxide 71
Barium nitrate 68
Beer's law 11
Bi-bivalent sulphates 42, 61, 63, 77–80, 93, 94
Bjerrum critical distance 92, 108
Bond resonance 88
Boric acid 88

Cadmium chloride 68
Caesium iodide 66, 99
Calcium chloride 66, 67, 68
Calcium hydroxide 41, 71, 94
Carbonic acid 88
Chelate stability 104
Chloroacetic acid 70
Classical equilibrium constant 2, **4**
Cobalt sulphate 16
Cobaltous oxalate 57, 59
Colligative properties 61
Colloidal electrolytes 30
Complex 6
Complexity constant 6
Conductance 25
Contact ion-pairs 7
Copper sulphate 16, 21, 32, 64, 78, 80
Corresponding solutions 18
Cryoscopy 61, 65

Crystal field stabilization 102
Cupric ammines 55
d^0 cations 97, 109
d^{10} cations 100
d^n cations 102

Dicarboxylic acids 85
Dielectric constant 107
Dihydrogen phosphate ion 38
Dimethylthallic hydroxide 22, 71
2, 4-Dinitrophenol 12
Dissociation constant 6
Distribution method 51, 55

Electrochemical cells 36
Electron-transfer reactions 37
Electrostatic model 90, 91, 94, 96, 100, 104, 107, 109
Entropy of reaction 83, 91
Equilibrium
  constants 4
  quotients 4
Ethylenediamine tetra-acetate ion 105
External ion-pairs 7
Extinction coefficient 11

Ferric ion, hydrolysis 4, 15
Ferric thiocyanate complexes 11, 14
Fluoroacetic acid 86
Formation constant 6
Free energy of reaction 83
Fuoss–Onsager equation 28, 31, 32, 34, 67, 108

Glass electrode 39

Heat of reaction 83, 91
Hydrochloric acid 36, 38, 44
Hydrogen chloride 87
Hydrogen chromate ion 15
Hydrogen fluoride 87
Hydrogen sulphate ion 40, 70

114 INDEX

Hydrogen sulphide 87
Hydrolytic equilibria 34
Hydroxylammonium ion 13

Inductive effects 86
Inner-sphere complex 7, 9, 10, 16, 79, 80, 93, 103
Inorganic oxyacids 87
Insoluble sulphides 95
Instability constant 6
Internal ion-pairs 7
Intimate ion-pairs 7
Iodic acid 88
Ion-exchange method 58
Ionic hydration 95
Ionic strength 5
Ionization constant 6
Ion-pair 6, 91
Irving–Williams order 103
Isopiestic measurements 65

Lanthanum cobalticyanide 32
Lanthanum ferricyanide 93
Lead chloride 42
Leist equation 28, 31
Ligand number 19, 46
Limiting law of Debye and Hückel 5
Lithium hydroxide 66
Localized hydrolysis 94, 100

Magnesium acetate 68
Magnesium sulphate 31, 33, 43, 61, 62, 64, 78
Manganous oxalate 31
Mesomeric effect 88
Mixed electrolytes 34, 42, 65
Mixed solvents 39, 107, 109
Molar absorptivity 11
Molar extinction coefficient 11

Nickel sulphate 9, 16
Nitric acid 20, 23
Nitromethane 88
Non-aqueous solvents 30, 39, 107
Nuclear magnetic resonance spectroscopy 9, 22

Onsager's limiting law 26
Optical absorption 11
  spectroscopy 9
Optical density 11

Osmotic coefficients 65
Outer-sphere complex 7, 16, 17, 80, 93, 94, 100
Overlapping complex ion equilibria 18, 46, 96
Oxyanions 98, 102

Partition function quotient 83, 91
Partition functions 82
Perchloric acid 23, 110
Phosphine 87
Polarographic method 48, 49, 72
Polybasic acids 87
Polynuclear complexes 48, 52, 95
Potassium chloride 107
Potassium dichromate 15
Potassium hexafluorophosphate 34, 67, 94
Potassium iodide 107
Potassium nitrate 66
Potassium sulphate 68
Precipitation 95
Primary salt effect 69
Protolytic equilibria 12, 19, 22, 25, 38, 55, 70, 77, 82, 85, 110

Raman spectroscopy 9, 19
Rates of reaction 69
Redox equilibria 90
Relaxation spectrometry 74

Silver ammines 47
Silver chloride 51
Silver nitrate 66
Sodium acetate 2
Sodium chloride 65
Solubility 51, 99
  equilibria 95
Solvent-separated ion-pairs 7
Solvent-shared ion-pairs 7
Specific interaction coefficient 6, 41, 44, 68
Specific solvation 109
Spectrophotometric method 9, 18
Stability constants 6, 90
  quotients 6, 46
Standard electrode potentials 37
Statistical effect 96
Step stability constants 96
Substituent effects 85
Successive stability constants 46
Sulphuric acid 20, 23, 65
Sulphurous acid 77, 88

INDEX

Swamping ionic media   3, 5, 44, 65

Tetra-alkylammonium salts   30, 99
Thallous acetate   66
Thallous chloride   31, 53
Thallous hydroxide   23, 71

Thermodynamic equilibrium constant   4, 15
Transmittance   11
Triphenylmethyl chloride   18
Water   39, 87

Zinc chloride   68